Energy Security in Asia

Energy security has become an increasingly important geopolitical issue, with concerns over soaring oil prices and surging Chinese and Indian energy imports compounded by the attempts of Chinese and Indian companies to purchase equity in oil production operations in the Middle East, Central Asia, Africa, Southeast Asia and Latin America. Many commentators have warned about the coming of 'resource wars' between Asia's rising powers and established, developed states. This book explores the various dimensions of energy security in Asia, examining the imperatives, dynamics and implications of Asia's rapidly expanding energy consumption and the growing need of East and South Asian countries to import energy at a time of rising global energy demand. It focuses on the challenges and imperatives facing the major players in the Asian energy security picture: the United States, Japan, China and India, as well as Asia's major energy producers: Russia, West Asia/Persian Gulf, Central Asia, and Australia. In each case, the domestic politics of energy security are investigated, and state interests and perspectives on the issue are considered. It goes on to analyse the policy and security aspects of energy security, including the geopolitics of energy competition; strategic, economic and environmental dimensions; and the impacts of energy security on human security.

Michael Wesley is the Director of the Griffith Asia Institute at Griffith University. He is the Research Convenor of the Australian Institute of International Affairs, a member of the Editorial Board of the *Australian Journal of International Affairs*, and a member of the Australian Research Council's College of Experts.

Routledge Security in Asia–Pacific Series

Series Editors:

Leszek Buszynski, *International University of Japan,*
and William Tow, *University of Queensland*

Security issues have become more prominent in the Asia–Pacific region because of the presence of global players, rising great powers, and confident middle powers, which intersect in complicated ways. This series puts forward important new work on key security issues in the region. It embraces the roles of the major actors, with defense policies and postures and their security interaction over the key issues of the region. It includes coverage of the United States, China, Japan, Russia, the Koreas, as well as the middle powers of ASEAN and South Asia. It also covers issues relating to environmental and economic security as well as transnational actors and regional groupings.

Energy Security in Asia

Edited by Michael Wesley

Routledge
Taylor & Francis Group

LONDON AND NEW YORK

First published 2007
by Routledge
2 Park Square, Milton Park, Abingdon, Oxon, OX14 4RN

Simultaneously published in the USA and Canada
by Routledge
711 Third Avenue, New York, NY 10017

Routledge is an imprint of the Taylor & Francis Group, an informa business

First issued in paperback 2012

Typeset in Times New Roman by Keyword Group Ltd

British Library Cataloguing in Publication Data
A catalogue record for this book is available from the British Library

Library of Congress Cataloging in Publication Data
A catalogue record for this book has been requested

ISBN13: 978-0-415-41006-9 hardback
ISBN13: 978-0-415-64748-9 paperback

Contents

Figures and tables

Figures

Tables

Contributors

Gawdat Bahgat is Professor of Political Science and Director of the Center for Middle Eastern Studies at Indiana University of Pennsylvania, USA. He is the author of several books including *Israel and the Persian Gulf* (2005) and *American Oil Diplomacy in the Persian Gulf and the Caspian Sea* (2003). His work has been translated to several foreign languages.

Stuart Harris is Emeritus Professor at the Department of International Relations at the Australian National University. Formerly Secretary of Australia's Department of Foreign Affairs and Trade, he has published widely on International Political Economy, Environmental Issues, and Foreign Policy.

Michael Heazle is an Australian Research Council Post-Doctoral Fellow with the Griffith Asia Institute. He is Author of *Scientific Uncertainty and the Politics of Whaling* (2006) and the Co-editor of *Beyond the Iraq war: the promises, pitfalls and perils of external interventionism* (2006) and *China–Japan Relations in the 21st Century: Creating a Future Past* (forthcoming).

Purnendra Jain is Professor of International Relations and Head of the Centre for Asian Studies at Adelaide University, Australia.

Aynsley Kellow is Professor at the School of Government, University of Tasmania, Australia, and a member of the Policy Program at the Antarctic Climate and Ecosystems Co-operative Research Centre.

Richard Leaver is Associate Professor in the School of Political and International Studies at Flinders University, Adelaide, Australia. He has published widely on Australian Foreign Policy and International Political Economy.

Satu Limaye is a member of the research staff in the Strategy, Forces and Resources Division of the Institute for Defense Analyses in the United States. He was formerly the Director of Research at the Asia–Pacific Centre for Security Studies in Honolulu.

Ashutosh Misra is a Research Fellow at the Institute for Defence Studies and Analyses, New Delhi. He holds a PhD on Indo–Pakistan Negotitations from

Jawaharlal Nehru University and is one of the leading Defence and Security Analysts in India.

Barry Naughten is a PhD scholar in the Centre for Arab & Islamic Studies (Middle East & Central Asia) at the Australian National University (ANU), Canberra.

William T. Tow is Professor of International Security in the Department of International Relations, Research School of Pacific & Asian Studies (RSPAS), The Australian National University. He is currently Editor of the *Australian Journal of International Affairs* and Co-editor of the Routledge Security in Asia–Pacific Series

Michael Wesley is Professor of International Relations and Director of the Griffith Asia Institute at Griffith University. Previously Assistant Director-General for Transnational Issues in the Australian government's Office of National Assessments, his most recent book is *The Howard Paradox: Australian Diplomacy in Asia, 1996–2006*.

Xu Yi-chong is Professor of Political Science at St. Francis Xavier University in Canada and Visiting Fellow at the Centre for Governance and Public Policy at Griffith University in Australia. Her most recent book is *The Governance of World Trade: International Civil Servants and the GATT/WTO*.

Preface

In 2005, oil prices on global markets reached $70 per barrel – in real terms, higher than those of the two oil shocks at the beginning and end of the 1970s. Unlike these previous shocks, which were the result of supply-side disruptions, the recent price rises have been driven by expanding demand, particularly in Asian developing countries. The two developing giants, China and India, are leading this surge; by 2025 it is estimated China's energy consumption will have grown by 156 per cent and India's by 152 per cent. Increasing attention has been directed towards Beijing's and New Delhi's attempts to secure access to oil and gas supplies in the Middle East, Africa, the Americas, and Southeast Asia. Because access to energy is essential to societal functioning, and the world is still dependant on finite stocks of fossil fuels, many commentators have warned about the coming of 'resource wars' between Asia's rising powers and established, developed states.

Studies on energy security tend to occur in transient waves and to comment narrowly on issues arising from the latest 'energy crisis'. The first oil shock in the early 1970s generated a surge of studies focusing on the world's heavy dependence on imported oil and its consequent vulnerability to manipulation by oil-producing countries (Inglis, 1975; Szyliowicz & O'Neill, 1975; Cook, 1976; Foley, 1976; Conant & Gold, 1978; Stobaugh & Yergin, 1979). The literature following the second oil shock later that decade focused on the interrelation of geopolitical instability and energy security and the environmental impact of fossil fuel use (Mabro, 1980; Smil & Knowland, 1980; Deese & Nye, 1981). A surge of studies followed the recent rise in oil prices, either predicting that unprecedented fossil fuel use is leading to the depletion of an exhaustible resource (Klare, 2004a; Roberts, 2004), or that an increasingly efficient international energy market will easily cope with surging global demand (Lomborg, 2001; Huber & Mill, 2005).

This book examines the imperatives, dynamics and implications of Asia's rapidly expanding energy consumption and the growing need of East and South Asian countries to import energy at a time of rising global energy demand. Part 1 focuses in on the challenges and imperatives facing the major players in the Asian energy security picture: the United States, Japan, China and India. Part 2 examines Asia's major energy producers: Russia, West Asia/Persian Gulf, Central Asia, and Australia. Part 3 provides a thematic approach to energy security in Asia, examining the issues from the geopolitical, strategic, economic, environmental and human security perspectives.

First draft chapters for this book were presented at a Workshop hosted by the Griffith Asia Institute, and funded by the Australian Research Council's Asia–Pacific Futures Network, in Brisbane, Australia, on 31 August – 2 September 2005. The editor and contributors are grateful to the staff of the Griffith Asia Institute and to Louise Edwards, the convenor of the Asia–Pacific Futures Network, for the support they provided to the Workshop and to the process of putting together this volume. Special thanks also goes to Will Alker and Jo Gilbert for their assistance with research, drafting, and helping to prepare the manuscript for publication.

Michael Wesley
Brisbane
7 April 2006

1 The geopolitics of energy security in Asia

Michael Wesley*

The break-up of the Soviet Union, the spread of radical Islam, and the rise of oil prices have led several authors to refer to a 'new great game' of great power competition in Asia (for example Arvanitopoulos, 1998; Rashid, 2001: Part 3; Klare, 2004). Whether the source of invading hordes or the site of Mackinder's (1950) 'geographical pivot of history', the Asian heartland has always exercised a powerful hold over popular conceptions of power and vulnerability. But leaving aside these dramatic images, there are two intriguing geopolitical trends occurring on the Asian continent, and energy security plays a central role in both. The first trend is that Asia's sub-regions are increasingly being integrated with each other through flows of commerce and lines of conflict in ways not seen for centuries. It will become less and less viable to think of Asia as a continent of separate sub-regions, whose links with the outside world are stronger than relations with other Asian sub-regions. A corollary to this is that Asia's commerce and conflicts have begun to engage outside powers with much greater urgency. The second trend is that despite this growing interdependence and the spread of the insistent logic of the market across Asia, strategic competition is growing among Asia's emerging great powers and with the United States.

Studying the geopolitics of energy security in Asia is important in its own right: without being too alarmist, there are dynamics afoot that could have a major impact on our security and prosperity in the near to medium term. But such a study also provides us with a unique window into the macro-patterns that will shape international relations into the future. The two geopolitical dynamics that I discuss in this paper provide intriguing insights into the likely patterns of US strategic involvement in Asia, and the reaction of Asia's rising great powers – whether allies, rivals, or non-aligned with Washington – to US policies. This, in turn, can tell us much about the future of the regional and global orders. My paper is divided into four parts. The first looks at the compelling logic of energy security provided by the 'anti-geopolitics' of the global energy market, but notes that many of Asia's great powers – energy importers and exporters alike – seem to be hedging against putting all of their faith in the market. The second part examines the first of two geopolitical dynamics in Asia, the maritime geopolitics of the sea-borne energy trade and the stability of the Persian Gulf sub-region. Part three concentrates on the territorial geopolitics centred on Central Asian energy producers and the routes of pipelines transporting oil and gas to prospective markets. My conclusion draws together the implications of each of these studies and speculates on the impact of Asia's geopolitics on the broader regional and global security orders.

The logic of market security

There are two important trends in the economics of the energy market as it relates to Asia. One is that Asia is producing, trading and consuming a growing proportion of the world's energy. In terms of consumption, the economic growth of East and South Asian economies will translate into a massive increase in demand for energy. The US Department of Energy estimates that China's consumption of oil will increase by 156 per cent by 2025, and that India's oil consumption will rise by 152 per cent over the same period. The proportion of oil and gas in the total energy mix of Asia's developing economies is projected to rise, and most major economies in North, Southeast and South Asia will rely on imports of oil and gas. In terms of production, the countries of West and Central Asia will provide an increasing proportion of global oil and gas production, as estimates of reserves predict that other sources will run out sooner and will be unable to provide surplus capacity (Porter, 2001).

The second trend is that the global energy trade is regionalising, due to the increasing transparency and globalisation of energy markets, which make transport costs and logistical considerations the major shapers of energy commerce (Odell, 1997). Thus, the major energy importers in East and South Asia are increasingly relying on West Asian exporters for their oil and gas, as North America moves towards Atlantic basin supplies and Europe towards Central Asian, African, and Atlantic Basin supplies (Salameh, 2003: 1087). In the words of the APEC Energy Research Centre, West Asia has emerged as the 'natural supplier' of Pacific Asia, which imports 80 per cent of its oil from West Asia. The Asia–Pacific takes 60 per cent of West Asia's oil and gas production and two-thirds of its exports (APEC Energy Research Centre, 2003: 47). These rising energy trade intensities between West Asia and East and South Asia are also being matched by interdependent investment patterns, with East Asian economies beginning to invest in upstream energy-producing operations in West Asia, and West Asian investors buying into downstream refining and distribution in East Asia (Salameh, 2001). Energy importers in South and East Asia are also showing increasing interest in Central Asian sources of oil and gas production. Although lagging behind the level of trade and investment with West Asia, the East and South Asian presence in Central Asian energy markets is growing (see Ogutcu, 2003; Daly, 2004; Blank, 2005). The energy trade demonstrates, perhaps more clearly than any other commodity, the growing integration among Asia's sub-regions.

Beneath the dynamics of increasing marketisation and regionalisation of energy in Asia lies an uncomfortable fact: as Asian societies become more dependent on imported energy, they become more vulnerable to irregularities of supply and affordability. Economists speak of energy's 'hysteresis effect', whereby modern economies have become so dependent on fossil fuels they are unable to do without them; and their ability to access them at an affordable price has come to be seen as an entitlement, and indeed as a facet of state security. Asia's developing economies, given their higher energy intensities, less diversified economies and higher proportion of manufacturing and agricultural industries, are especially vulnerable to energy supply disruptions or price rises. But despite these vulnerabilities, the long-term trend over several decades has been to move away from exclusive, long-term statist arrangements for energy supply towards open global energy markets.

(May, 1998: 16–17)

The historical movement has been towards the 'anti-geopolitics' of the market, in which it is market presence and position, and the ability to secure contracts, rather than territorial access or state-to-state deals, that delivers energy security. Cartels have lost their ability to manipulate prices after the 'reverse oil shock' of the 1980s, a loss of market power due to a combination of conservation efforts, the development of alternative energy sources and the discovery of new oil fields. Prices are no longer determined by deals between producers and distributors, but rather by spot markets and futures contracts negotiated openly and competitively (Fesharaki, 1999: 91)

To a pure economic logic, there is no contradiction here. The increasingly free market in energy, when backed up by measures such as strategic petroleum reserves and regional crisis sharing arrangements,[1] provides security for importing economies in three ways. First, the energy market diversifies risk and provides the capacity to absorb disruptions through the price effect. Fossil fuels are a fungible commodity, and in the conditions of an open market, all consumers absorb an equal part of a supply disruption through paying the same, increased price. As Robert Manning argues,

> Insofar as oil is a globalised commodity, a disruption anywhere is a price spike everywhere. Thus, mere geopolitical access to the strategic resource will not yield the accessing party the best price. What matters, rather, is who gets what long-term contract.
>
> (Manning, 2000: 7).

The recent track record of the global energy economy in providing such security is impressive. Since 2002, the market has absorbed the effects of three major supply disruptions due to instability in Venezuela and Nigeria and the war in Iraq, with some help from demand weakness and restraint in major importers and a willingness of Saudi Arabia to boost capacity.

Second, the experience of the 'reverse oil shock' and the dependence of most oil exporters on socially determined target incomes from energy exports has led to a significant convergence of interests between producers and consumers in maintaining affordability and stability in global energy markets, and in the ongoing affordability of production and consumption. Energy exporters, while happy to enjoy the increased rents of temporary price spikes, worry about the effect of long-term price rises on the diversification of energy sources and demand restraint. This convergence of producer and consumer incentives complements the risk-spreading mechanism of the global energy market, making major producers such as Saudi Arabia more willing to return to the traditional role of 'swing producer' to flatten price fluctuations.

The third security mechanism provided by the global energy market derives from the growing interdependence among national economies, and between the energy market and global trade in other commodities and services. As Peter May observes, 'Asia depends for at least some and probably most of its continued economic growth on the continued normal functioning of world markets both for petroleum products and for the exports needed to pay for those products' (May, 1998: 25). In other words, economies in Asia require energy to fuel their export-led development, which in the main is provided by growing trade intensities with European, North American and other Asian economies. An energy embargo against any of these major economies would cause it to stall, with inevitable knock-on effects for other major economies. Thus, under current conditions of economic interdependence, it is impossible for any major economy to energy embargo another without itself suffering major economic losses. And as Edward Porter points out, sanctions against producers under conditions of tightening energy markets will almost inevitably be circumvented by other consumers, with the attendant price-effects probably offsetting the producer's losses from the original sanctions (Porter, 2001: 2).

Despite the historical trend towards the anti-geopolitics of the market as the provider of energy security, it is clear that major energy consumers and producers in Asia are developing alternative energy security policies. These alternative policies represent a statist quest for direct control over energy sources and supply routes, the opposite logic to a fuller integration into the market. In other words, two competing, geopolitical logics are asserting themselves. Japan, China, the Republic of Korea, and India have all embarked on programs to invest directly in fossil fuel production regions through state-owned or semi-state-controlled energy companies. Russia, Japan, China, India, Pakistan, Iran, Turkey and various Central Asian states have engaged in protracted competition over the routes of pipelines transporting oil and gas to export markets. To be sure, acquiring 'equity oil' does have some advantages. It reduces market risk by allowing an investor to predict accurately the amount of fuel received over the life of the field, and promises cheaper fuel through transfer pricing (Downs, 2004: 35). It provides a sense of psychological security against price fluctuations and supply disruptions, although experience suggests that equity oil's marginal share of the global oil market means it will have a limited impact on either price volatility or compensating for supply disruptions (Wayne, 2005: 5). On the other hand much of the investment in equity oil has been criticised for distorting energy investment flows and decreasing the efficiency of the financing of exploration and production. There is also evidence that competition for equity oil has developed its own zero-sum logic among Asia's consuming powers. For example, Indian Prime Minister

Manmohan Singh (cited in Klare, 2005b) is reported to have said at the start of 2005 to Indian energy companies, 'I find China ahead of us in planning for the future in the field of energy security. We can no longer be complacent and must learn to think strategically, to think ahead, and to act swiftly and decisively.' Is this the logical response of former command economies still uncomfortable with relying fully on markets? Or are Asia's emerging powers viewing energy within the context of a broader geopolitical competition currently being played out?

Asia's maritime geopolitics

In deference to Alfred Thayer Mahan (1890), it is possible to identify two distinct, though not competing, geopolitical logics in Asia: one maritime, one terrestrial. The maritime dynamic has been evolving for several decades, and is based on established patterns of US dominance and power projection and on mature institutions and processes. The terrestrial geopolitics are much more recent, though drawing on centuries-old logics, and are based on rising powers, recent political changes, and emerging structures and institutions. Both of these dynamics are strongly influenced by energy security concerns: access to energy supplies and markets, the integrity of delivery systems, and the minimisation of vulnerabilities to disruptions. In each geopolitical dynamic, major powers' perceptions of their geographic vulnerabilities, buffers and entitlements play a crucial role. In each, there is a tendency for strategists to discount current benign conditions and the efficiency of energy markets in light of possible future vulnerabilities. To paraphrase Robert Strausz-Hupe (1942: 7), geopolitics represents the state's geographic consciousness; in the current period it is intriguing to watch historic patterns, perceptions and liabilities shape current and future interactions.

The vast majority of the world's energy trade is sea borne. The United States, like the United Kingdom before it, regards its role in providing security for global maritime routes, and in particular the Persian Gulf, as an intrinsic part of its stewardship of the global economy. The United States' strategic purpose in West Asia is less to capture the region's energy resources for itself, as the conspiracy theorists suggest, than to prevent the development of destabilising competition in the region among other great powers. Walter Russell Mead (2004: 43) describes US strategic goals thus:

> The United States is less interested in feeding its oil thirst and in gaining contracts for powerful energy-sector companies than it is in the impact of oil security – or insecurity – on world politics as a whole. Because the United States has both the power and the will to maintain the security of the world oil trade, other countries see no adequate reason to develop their own independent military capabilities to secure their oil supplies. A world with half a dozen powers duelling for influence in the Middle East, with each power possessing the will and the ability to intervene with military force in this explosive region, would be a less safe and less happy world than the one we now live in, and not only for Americans.

Maintaining its position as the unrivalled guarantor of the maritime energy trade provides a powerful rationale for the general US strategic purpose of ensuring peace through overwhelming strength as set out in the 2002 National Security Strategy and promoting 'a balance of power that favours freedom' in its 2006 version (*National Security Strategy of the United States of America*, 2002; 2006). This policy also has the benefit of playing a crucial part in the United States' ability to maintain that hegemony, as US control over energy supply routes will curtail the possibility of the rise of a strategic competitor. According to one commentator, 'In this respect, control of oil may be seen as the centre of gravity of US economic hegemony and thus the logical complement of its declared strategy of permanent, unilateral military supremacy' (Bromley, 2005: 227).

The internal logic of US policy towards the world energy trade is to maintain its position as the sole guarantor of global energy transit, and thus to bring all energy supply routes ultimately under its security umbrella. As the main guarantor and advocate of a free global energy market, its objective is to bring all producers and consumers into that market, on conditions ensuring they comply with the requirements of the market. The other, crucial strand of US energy security policy is to extend the conditions of its maritime dominance inland in West Asia, the region containing 63 per cent of the world's oil reserves. Historically, the United States has tried to achieve this through a mixture of supporting key allies and containing or restraining states unwilling to comply with the projected US order in the region. For many years, US policy in West Asia was based on the 'twin pillars' policy of supporting Saudi Arabia and Iran as the guarantors of regional order. After the Iranian revolution and with Iraq becoming increasingly belligerent, the 'twin pillars' strategy was replaced by a policy of 'dual containment' of Iraq and Iran. Notably both strategies have been unstable and relatively short-lived. It remains to be seen how US strategy will evolve following the 2003 invasion of Iraq, but there is little reason to suggest that it will abandon its traditional objectives: promoting a regional order most conducive to the operation of a free market in energy; retaining its role as the sole guarantor of the global energy trade; and seeking to isolate and change states opposed to the United States' role and preferred sub-regional order.

At first blush, the US sole guarantee of maritime energy security offers many advantages to other states. With the US insisting that it shoulder alone the costs of providing this security, the world's energy importers are in effect being invited to free-ride on a public good paid for by the US taxpayer (Manning, 2005: 7). Yet there are signs that several Asian states – including allies of the US – are uncomfortable with this arrangement. A closer examination of US policy in West Asia suggests why this is the case. First, several Asian states have begun to worry about the effects of US policy commitments on regional stability in West Asia. According to Simon Bromley, in recent decades US policy in West Asia was 'based on a series of contradictory commitments that increasingly undermined its ability to play a directive role in the politics and geopolitics of the region' (Bromley, 2005: 248). In the context of US policy in relation to Iraq, the Israeli–Palestinian dispute and rising anti-Americanism, several Asian powers have begun to worry whether by acquiescing to the US energy security umbrella, they are leaving themselves vulnerable to collateral damage arising from Arab anger at US policies. This sentiment is of a kind with that reportedly expressed by a Japanese businessman during the Iranian revolution:

> Why is it that we who have had nothing to do with the causes of the Iranian revolution, nothing to do historically with the Arab–Israel conflict, and nothing to do with American interests in Iran, have to suffer this?
>
> (Cited in Dowty, 2000: 2)

Second, potential great power rivals realise that while the US promotes a free energy market now, this position could be used in future to pressure or contain potential rivals. Threatening energy supplies constitutes one of the most effective contemporary international mechanisms for following Sun Tzu's advice to 'subdue the enemy without fighting'. History shows that the US has been more than willing to use energy embargoes in this way. Its famous petroleum blockade of Japan in July 1941 was the result of a policy designed to contain Japan's expansionist ambitions without committing to war in Asia (Feis, 1950). Washington's oil embargo on Britain and France during the Suez crisis demonstrated it is willing to use this weapon against allies as well as competitors. Recent history shows that Washington has used its position of dominance over maritime energy flows to try to secure other policy goals. Its attempts to impose energy embargoes on 'rogue states' such as Iraq, Iran and Libya are examples not only of the extent to which the US values its position of dominance, but also of the extent to which that position is vulnerable to circumvention by other energy-hungry great powers.

Third, states such as China and Russia have every reason to dislike the prospect of working within a world determined solely by US guarantees, US-designed market arrangements, and US-sponsored institutions. Both states have experienced during the 1990s Washington's manipulation of their access to global and regional institutions as methods of extracting concessions and demanding structural adjustments. China's long road to accession to the WTO is a case in point; Russia's treatment in relation to NATO and APEC (where it was persuaded to accede to US policy on NATO expansion in exchange for admission to APEC) is another. While both states' interests are currently served by the operation of international energy markets, they and other Asian powers have reason to worry about how a US-determined and guaranteed global market may be used to discriminate against or contain them in the future (Katsuhiko, 2000).

The growing integration between West Asian and East and South Asian energy markets and the high vulnerability of many developing Asian economies to energy price fluctuations has led Asia's rising great powers – China, Japan and India – to realise they have a significant stake in the political order in West Asia. China's strategists worry about vulnerabilities to a US energy embargo in the event of Sino–US tensions as much as they worry about the effect of energy price rises on China's economic growth and therefore domestic stability (Downs,

2004). As a result, Chinese energy companies have launched a campaign of buying equity in oil and gas production operations in West Asia, North and West
Africa, and Latin America, and Beijing has been building diplomatic links with
West Asian states (see Dannreuther, 2003; Ogutcu, 2003; Daly, 2004; Bajpaee,
2005). China has paid particular attention to forging close political and commercial links with West Asian regimes isolated by the US and thus outside of the US-
sponsored regional and market orders, such as Saddam's Iraq and Iran
(Feigenbaum, 1999).

Given its dependence on West Asian energy imports has climbed to above
1973 levels, Japan also has become more proactive in its involvement in West
Asia. Tokyo worries about its vulnerabilities to supply disruptions, and on the
effects an oil shortage may have on China's strategic behaviour (Myers-Jaffe &
Manning, 2001: 144). Given its close alliance with the US, Japan has a stake in
trying to moderate US policy towards West Asia (Lesbirel, 2004). In recent years,
Japan has begun to build 'extended security' relations with several West Asian
states (Calabrese, 2002), particularly concentrating on peace, governance and
stability-building assistance (Dowty, 2000), in a clear effort to differentiate itself
from US policy and the region, but without cutting across US interests there. Of
particular note has been Japan's willingness to displease Washington in developing closer commercial and diplomatic links with Iran (Calabrese, 2002: 93).

India's interests in West Asia are both commercial and strategic. Its growing
demand for fossil fuels and proximity to West Asia dictate a keen economic interest in the region; its intention to play the role of a maritime power in the Indian
Ocean increases its stakes especially in the Persian Gulf (Blank, 2004). India's
interests in combating terrorism and isolating Pakistan have also drawn it diplomatically towards West and Central Asia (Parthasarathy & Kurian, 2002). In
recent years, India has intensified its diplomatic, economic and defence linkages
with West and Central Asian states. In addition to the quest for equity oil and gas,
India signed a defence co-operation agreement with Iran in 2003 and negotiated
an air base deal with Tajikistan. New Delhi has also begun to realise that these
links have begun to translate into diplomatic coin in Washington. During her
March 2005 visit to New Delhi, US Secretary of State Condoleeza Rice urged
India to abandon a major deal to import natural gas overland from Iran. The
February 2006 deal between Washington and New Delhi on civil nuclear co-operation was interpreted by many as being motivated in part by a US desire to
remove an energy imperative from the growing closeness of Iran–India relations.

The picture that emerges here is that Asia's rising great powers – whether
allied to the US, potential rivals, or non-aligned – are hedging against the US
advocated and guaranteed West Asian order. Other regional powers, such as
Russia and Iran, also have a major stake in West Asian developments. And while
West Asian states such as Saudi Arabia for many years have been willing to buy
US-guaranteed military security in exchange for their agreement to co-operate
with US demands on energy market pricing and production as well as the sub-
regional order, US demands after 9/11 may be becoming too onerous. US allies
in the region may start to become more open to alternative suitors. The maritime

geopolitics of energy security in Asia have become the site of vigorous great power positioning; whether this positioning develops into outright competition depends in large part on the perceived ramifications of US policy in West Asia in the coming years.

Asia's terrestrial geopolitics

The vast Central Asian plain is a region without many natural frontiers, a geographic fact that has made Asia's heartland historically the site of power spillovers and territorial competition. Abutting this region are China, Russia, Iran and Turkey, which in Goblet's terms are 'extensive' states – territorially large, exploitive of human and physical resources and profligate in their use, and given to territorial expansion as much as they expect it from others (Goblet, 1955). Each of these four states feels deeply vulnerable in relation to its Central Asian borders and regions. China's two most restive provinces, Tibet and Xinjiang, are also its western-most. Beijing particularly worries about pan-Turkic sentiment and Islamic radicalism crossing its porous borders into Xinjiang and stoking centripetal tendencies. Russia is concerned about the slow depopulation of eastern Siberia and the growing Sinicisation of the region. Eastern Siberia, like Xinjiang, contains significant fossil fuel deposits. Iran and Turkey also worry about the porousness of their borders to damaging transnational flows, from terrorism to drugs. Both states are also worried about the irredentist urges of minorities – particularly Kurds – on their territorial peripheries. Given these vulnerabilities, the historical tendency has been towards a 'defensive expansionism': Russia and China in Central Asia; Russia, Turkey and Iran in the Caucasus; and India, China and Pakistan on the Himalayan elevation. More recently, these competitive urges, while still present, have been moderated by co-operative ventures. Realising their mutual vulnerabilities, and the damage that could be done if they began stoking insurgencies in each other's territories, the Asian great powers have forged common cause against separatism. Through mechanisms such as the Shanghai Co-operation Organisation (SCO), China and Russia have sought to offset their territorial vulnerabilities. India, Pakistan and Iran attended their first meeting of the SCO as observers in October 2005 (SCO n.d.).

Energy adds complexity to the dynamics of great power competition and collaboration in Asia. Post-Soviet Central Asia has emerged as a region with significant fossil fuel reserves as well as a possible alternative to maritime transport routes between West Asia and East and South Asia. The terrestrial order therefore offers an alternative to the US-dominated international energy trade. Thus the other power with an interest in the Central Asian order is the US. Realising that energy sourced and transiting through Central Asia has the potential to erode its position as the sole supplier of global energy security, Washington has established a significant presence in the region. Its expanded diplomatic links were complemented after 9/11 by several basing arrangements in Central Asia, developments which greatly increased China's, Iran's and Russia's geostrategic discomfort, and led to renewed rivalry in the region.

The final element in a volatile mix in Central Asia is supplied by the regimes of the post-Soviet Central Asian states. Most have emerged as secular authoritarian governments, plagued by domestic instability and territorial disputes, subject to great power (mainly Russian) interference, and threatened by growing Islamist grassroots movements. For those with significant oil gas reserves, these resources simply exacerbate many of these problems, by heightening territorial competition among states and increasing domestic resentment over the inequitable distribution of oil and gas rents. In many Central Asian states, the patterns and dilemmas are reminiscent of those in West Asia: the choice between authoritarianism and theocratic populism, the growing urgency felt by regimes to capture oil and gas rents, and the increasing stakes of outside powers in the domestic regime makeup of the states.

The resulting geopolitical competition is different from its maritime counterpart. In the first place, while the US is a major player, it is by no means the main influence on the Central Asian order. For Washington, energy complements and draws together its other main strategic interests in Central Asia: terrorism and proliferation (Klare, 2005a). Its main objectives in Central Asia are to draw the region's regimes into the Western orbit, particularly in relation to the energy trade and the War on Terror, to intensify Central Asia's economic links with Western Europe, to isolate Iran and to curb Russian and Chinese influence (Myers-Jaffe & Manning, 2001: 143). The US wants to lock in Central Asia's energy supplies as an integral part of a diversified single global energy market.

Most of these goals put Washington in direct competition with Russia and China. Moscow sees the region as part of its traditional sphere of influence and is deeply opposed to interloping great powers. Energy transfers and security assistance have been two of its main levers for exercising influence over its 'near abroad'. Russia has several objectives in seeking to gain influence over Central Asia's energy exports. Russian nationalists cite the Soviet role in developing oil and gas fields in Central Asia and the Caucasus, and demand a cut for Russia of all subsequent production. More dispassionately, Moscow has sought to influence the routing of pipelines through its territory as a way of offsetting its own vulnerabilities, as its western distribution of Russian energy production runs through the territory of various US allies.

In addition to realising their mutual vulnerabilities, Russia and China have also found common cause in limiting US influence in Central Asia. Beijing realises the potential for Russian and Central Asian energy to relieve its dependence on maritime supplies. In recent years, Russian and Chinese policy towards Central Asia has increasingly been co-ordinated into a joint strategy for limiting US influence through the Shanghai Co-operation Organisation (SCO), an organisation linking Russia and China with Kazakhstan, Uzbekistan, Tajikistan and Kyrgyzstan. The SCO has developed its own counter-terrorism training and joint exercises and has promoted economic integration among its members – each component of which seeks to counter a US objective. Moscow and Beijing are unambiguous in their support for Central Asia's authoritarian regimes in their struggles against separatism and Islamist populism. This seems to have provided an edge over the US, whose support for democratic reform in the region left it vulnerable to a Sino–Russian campaign seeking to dislodge US bases from Uzbekistan and Kyrgyzstan in mid-2005.

Iran has been a notable beneficiary of Asia's terrestrial geopolitics. Its own substantial, US-embargoed energy reserves, plus its geographic position as a potential strategic 'hub' linking Central Asian energy fields with maritime distribution through the Persian Gulf, have enhanced its strategic coin substantially. Not only have China and India negotiated major energy supply deals with Tehran (with China by the end of 2004 becoming Iran's largest export market), but so have significant US allies such as Japan and Turkey.[2] In an era of rising oil prices and vanishing surplus capacity, foreign companies are undeterred by the United States' Iran–Libya Sanctions Act (ILSA) which threatens embargoes against companies doing business with Iran; indeed some have been willing to challenge the WTO-compatibility of ILSA. Tehran seems to have been successful in converting some of this strategic coin into diplomatic coin, and can likely rely on a Chinese and possibly a Russian veto as the issue of Iran's nuclear program is debated in the UN Security Council. Thus Iran, Washington's major bete noir in West Asia, is inarguably its strongest geopolitical situation in decades. The invasions of Afghanistan and Iraq, and the Syrian withdrawal from Lebanon, have provided it with unprecedented regional leverage. Its traditional Arab rivals, Ba'athist Iraq and Saudi Arabia, have either been eclipsed or distracted by internal turmoil. Its growing links to Russia, China and India are increasingly subverting US-backed isolation, and Tehran now has a range of deterrents against direct US attack either in hand or very close to hand: terrorism in the Levant; order and disorder in Afghanistan and Iraq; and closing in on a nuclear weapon.

Conclusion

The geopolitics of energy security in Asia present a unique perspective into some emerging trends in international relations. Influencing the conditions of production and supply in Asia are central to the United States' global role, but these tasks are beset by major challenges. The combination of rising oil prices, vanishing surplus capacity, growing concerns over energy security, and competing US policy commitments in West and Central Asia constitute major obstacles to Washington's objectives. While on the surface, the US-guaranteed global energy market is working efficiently, underneath there are signs that both US allies and rivals are hedging against the market and the US-determined order in West Asia. One commentator suggests that if present trends towards the competition for equity oil continue, the United States itself may be forced towards a more mercantilist approach to the security of supply (Bromley, 2005: 255). Faced with these obstacles, the US may slowly lose its role as sole guarantor of the world energy trade and its ability to determine the regional order in West Asia.

If this is the implication of emerging trends in energy security in Asia, it has major implications for regional and global security. An Asia in which US policy is increasingly challenged and subverted by emerging great powers would seem to portend a period of regional and global instability. The vision summoned by Walter Russell Mead, of 'a world with half a dozen powers duelling for influence in the Middle East, with each power possessing the will and the ability to intervene with military force in this explosive region' (Mead, 2004: 43), is an extremely dangerous and pessimistic one. Yet this need not be the only outcome of current geopolitical trends. One alternative, already trialled in Central Asia, is that powers with partly competitive but partly mutual interests will forge a cooperative arrangement for managing the regional order. In order for this to happen, each of the stakeholders will need to recognise the direction of the geopolitical trends and respond rationally to the options they present.

Notes

* My thanks to Will Alker for his research assistance in preparing this paper.
1 Such as the ASEAN Petroleum Security Agreement (APSA).
2 In 1996, Turkey signed a $3 billion 25-year deal to access Iranian natural gas.

Part I
Demand

2 The United States and energy security in the Asia–Pacific

*Satu P. Limaye**

The cover of a recent issue of the *Economist* showed caricatures of the United States and China under the headline, 'the oilaholics'. While the view that the US and China are in direct competition for scarce energy resources in the region is popular in non-official circles, it does not represent the major concerns of the US government. This chapter will argue that for the US, energy security issues in the Asia–Pacific are less concerned with where pipelines are built or the emergence of competition over energy resources, than with the management of energy needs and issues in the context of Washington's wider strategic approach to the region. Fundamentally, energy security issues in the Asia–Pacific do not play a major role in US policy and are only considered within a broader geopolitical, strategic framework.

The first purpose of this chapter is to discuss why energy security in the Asia–Pacific has so little significance for US policy. More important, from Washington's perspective, are western hemisphere suppliers, specifically Mexico, Venezuela, and Canada. West Africa is growing in importance, and the Middle East continues to be vital for the stability of the world market and supply to key allies in Europe and Japan. Secondly this chapter will compare the official and unofficial responses to energy security issues in the Asia–Pacific. The official response, or lack thereof, is reflective of the relative unimportance of the region in US energy security calculations. The unofficial debate may reflect a geo-strategic discourse that debates the rise of China and the growing importance of Central Asia but does not reflect the realities of the US approach to energy security in the region. Thirdly this chapter will look at the energy security needs of Asia–Pacific states and how the recalibrating of their policies may affect the US. It is important to note that states will pursue their own national interests in regard to energy security and these actions may infringe on US interests. In the broad scope of US alliances in the region, energy security issues are unlikely to have a dramatic effect on Washington's key relationships, but they may affect the outcome of US strategic goals. Finally this chapter will look at the geo-strategic issues in the Asia–Pacific that are exacerbated by energy security concerns. For the US the major energy security issues in the Asia–Pacific have little to do with the use of energy and a great deal to do with the political consequences of energy use.

US energy security interests: weighing the Asia–Pacific's importance

For the US, energy security and the Asia–Pacific are usually considered mutually exclusive due to Asia's lack of resource volume in US energy calculations. US perceptions about the nature and seriousness of energy competition in the Asia–Pacific are taking shape as regional events unfold. Other regions of the world currently receive much more attention within the US if the issue is framed more generally as one of energy security. As a result discussions about energy security in the US rarely refer to the Asia–Pacific. The reason for this is two-fold. First, US dependence on the Asia–Pacific region for its energy supplies is negligible. Second, the United States' concerns about energy security are far more urgent and immediate in the case of the western hemisphere, Africa and the Middle East for the simple reason that the bulk of US energy imports originate from the western hemisphere and Africa; and because the Middle East remains an important source of energy for key allies in Europe and the Asia–Pacific. The top 10 suppliers of US oil imports in 2002 were, in order, Canada, Saudi Arabia, Venezuela, Mexico, Nigeria, Iraq, Colombia, Norway, Britain, and Angola. Not one Asia–Pacific country makes the top ten on the list. While the rankings have changed somewhat over the years, and a few new sources of energy have been added, it is noteworthy that the composition of countries upon which the US relies for its energy has not changed fundamentally in nearly half a century:

> Since the early 1960s, the United States has been steadily increasing its reliance on imported supplies of both petroleum and natural gas. Whereas in 1960 the country imported about 17 percent of its oil, by 2002 imports accounted for nearly 53 percent of total oil use. The largest suppliers of U.S. imports have changed somewhat over this period, along with their relative importance. In 1960, Venezuela, Canada, Saudi Arabia, Colombia, and Iraq were the largest suppliers of foreign oil to the United States. In 2002, Canada provided the largest share of U.S. imports, followed by Saudi Arabia, Mexico, Venezuela, and Nigeria. In addition, the number of oil exporters to the United States has increased, with supplies coming from Angola, Argentina, Ecuador, Norway, and the United Kingdom, among others.
>
> (Caruso & Doman, 2004)

This is not the case, for example, for China. Thus US interest in energy security in the Asia–Pacific will not arise from the usual supply and demand scenarios because it plays such a minor role in US energy supplies. This is not to suggest that the Asia–Pacific doesn't hold a strong position in geopolitical thinking but rather that it has little relevance to the United States' own energy security calculations.

Indeed, the point could be made that as a matter of national security, energy-related developments in the Asia–Pacific should take a back seat to developments in the western hemisphere and Africa – specifically West Africa. According to the

Secretary of Energy Spencer Abraham (2004) '[e]nergy from Africa plays an increasingly important role in US energy security, accounting for more than 10 percent of US oil imports, and it is a key economic engine for the continent.' This amount is expected to grow appreciably in the coming decades. An indication of the public resonance in the US of African energy developments is a recent three-part series on energy production in troubled Nigeria broadcast on US National Public Radio entitled 'Oil Money Divides Nigeria' (2005). As US energy imports from Africa increase and concerns rise that an African continent is giving way to chaos and becoming an 'ill-governed space' ripe for terrorist activities, the US is likely to focus on getting the energy and security picture right on that continent. Moreover, the future of Africa will continue to attract domestic political attention and the energy issue may well provide a 'hook' that has previously been lacking. This will aid a number of US strategic interests, such as reducing poverty, fighting terrorism and meeting increased energy demands (*National Security Strategy of the United States of America*, 2002: 4).

Closer to home, the western hemisphere is even more critical to US energy interests. Fifty-two per cent of US crude oil imports and 54 per cent of petroleum product imports come from the western hemisphere, where three of the top five exporters to the United States (Canada, Mexico and Venezuela) are located (Brodman, 2005). Venezuela, which possesses one of the largest reserves of oil outside the Middle East, has become a country of considerable concern given its rank as the third largest exporter of oil to the US and ongoing political developments there. The Reverend Pat Robertson's recent comment that:

> If he [President Hugo Chavez] thinks we're trying to assassinate him, I think that we really ought to go ahead and do it. It's a whole lot cheaper than starting a war. And I don't think any oil shipments will stop

shows what a neuralgic issue energy security in the western hemisphere is within the US (Goodstein, 2005). And of course Mexico (the fourth largest source of US oil imports) and Columbia (the seventh largest source of oil imports) have been and will remain critically important US foreign policy priorities for a variety of reasons, including energy security. Canada is not only America's largest single supplier of energy, but also its largest overall trading partner. Further evidence of the importance of the western hemisphere for US energy security was demonstrated by the devastation and subsequent energy price surge following hurricane Katrina. Disruptions of US energy supplies from the western hemisphere have had a greater influence on fuel prices than any expanded demand from China or the billions of dollars spent on pipelines through Central Asia.

West Asia enters US energy security calculations in a slightly different way. Although it is true that though Saudi Arabia remains the second largest source of US oil imports, the region as a whole only accounts for just under 20 per cent of US oil imports. However, the importance of the region derives from two other facts. First, it is a critically important source of energy for US partners and allies – especially Japan and Europe. And second, the scale of production from the region helps to determine global prices. The importance of the Middle East was clearly demonstrated by the OPEC oil crises of the 1970s, where an artificial increase in world oil prices led to a slowdown of the world economy. While the US is considerably less exposed to this kind of risk and OPEC seems less willing to use oil prices to hold the world economy hostage, the Middle East remains essential for the stability of world oil prices. This is best demonstrated by Saudi Arabia's ability to increase supply to take up the slack during times of massive supply disruption in other countries. While Indonesian and Central Asian pipelines may be increasing in importance for their own economies and within the world energy supply equation, they remain remarkably unimportant in regards to US energy supplies.

US official and unofficial attention to energy issues in the Asia–Pacific

As established in the last section, in traditional energy security terms, the Asia–Pacific is largely irrelevant to the US, but energy security as an issue has a tendency to combine with and complicate geopolitical considerations. There has been a plethora of press and scholarly articles on the energy security situation in the Asia–Pacific region. Most commentary highlights the rapid increase in demand of rising economies such as China and India and the pressure that this is placing on world oil markets. However, these seem to have merged with a larger debate about economic competition, environmental concerns, and wider hegemonic stability questions. Washington's policies in this area have attracted little attention, but the US is currently engaged in a host of Asia–Pacific energy security activities. For example, there has been only one Congressional hearing (in the Senate) on the subject of the implications of Asian energy trends for the US and my discussions with House staff indicate that there are no such general hearings planned for the future. Nevertheless, the Congressionally established United States–China Economic and Security Review Commission regularly reviews 'the national security implications of trade and economic ties between the United States and the People's Republic of China, including an assessment of China's energy needs and strategies' (Chanlett–Avery, 2005: 1).

The new Commander of US Pacific Command, Admiral Fallon, during his recent statements to the House Armed Services Committee, made no explicit mention of Asia–Pacific energy competition in his prepared remarks. Nor did Admiral Fallon explicitly address the issue during his speech at the Shangri-la Dialogue. US Assistant Secretary of State for East Asia and the Pacific Christopher Hill, has not, to the best of my knowledge, made any specific mention of this issue in his public speeches or statements. President Bush's May 2001 National Energy Plan (NEPD) did make some suggestions that could be applied to Asia. However, it is worth noting that the Plan's most important recommendations on the international scene related to Africa, the western hemisphere and the Middle East. The only country in the Asia–Pacific region specifically mentioned was India. The NEPD called on the President to 'direct the Secretaries of State and Energy to work with India's Ministry of Petroleum and Natural Gas to help India maximize its domestic oil and gas production' (*National Energy Policy*, 2001). The Central Intelligence Agency's 2020 assessment has referred to the important impact that energy security issues will have on China's and India's foreign and security policies. All of this is not to say that energy competition and energy security issues in the Asia–Pacific are not of interest or concern to US policymakers, but it is to put these perceptions into perspective. Indeed, a central argument of this chapter is that the US policy challenge in the Asia–Pacific has less to do with which pipelines are built, the impact of energy needs on foreign and security policies of regional great powers, or energy resource wars and more to do with the management of energy needs and issues in the context of a wider strategic approach to the region and Washington's relationships there.

Major differences in official and non-official perceptions

The divergence of official and non-official perceptions of energy security issues in the Asia–Pacific again highlights the politicizing effect that energy security has. This of course does not mean that there are not overlaps, but rather that the official position is geared towards maintaining stability and managing the system, while the non-official position looks to highlight the possible pitfalls. First, US officials have sought to portray a more balanced, less breathless picture of the evolving energy situation in the Asia–Pacific. For example, they have noted in open testimony to Congress that it is important to keep in perspective the scale of growing energy demands in Asia – primarily of China and India. China's and India's economies, at 12 per cent and 5 per cent, respectively, are dwarfed by the size of the US economy. While each American consumes 28 barrels of oil per year, each Chinese consumes two barrels and each Indian less than one. Moreover, future demands for energy are uncertain and will depend upon continued economic expansion (Wayne, 2005). Second, US officials seem less concerned than their non-official counterparts about recent Chinese and Indian national oil companies purchasing overseas assets to develop direct access to supplies. For example, one non-government analyst advised that 'US policymakers need to find ways to discourage China and India from seeking to "lock-up" global equity oil supplies in a futile, mercantilist effort to monopolize those supplies for their own economies, i.e. "take oil off the market"' (Herberg, 2005a: 16). US officials do not deny that China and India are engaged in such a strategy, but have questioned whether such a strategy really does afford protection from the volatility of the market either in terms of supply or price. Moreover, they have noted that the prices paid for certain energy assets by China and India are well above what private sector western companies would contemplate. They note that Japan engaged in a similar strategy after the 1970s oil shocks only to abandon it due to poor results. However, both US officials and non-government analysts have expressed concern about the particular countries with which China and India are seeking to cooperate – namely, Iran, Myanmar, and Sudan – because of the impact it has on US policy efforts.

A third difference between the official and non-official perceptions is that US government officials have taken pains to stress the commonalities and emerging bases of cooperation rather than the prospects for outright competition or even conflict. Undersecretary of Energy David Garman (2005) told a Senate Committee that 'Rising demand has left major consuming countries such as the United States, China, and India with the shared goals of diversifying and expanding the oil supply sources available to the world market'. He went on to say, 'We still believe that cooperation with China, India and other major developing countries can bring US quicker and better solutions'. The official approach seems to be focussed on developing a more harmonious world energy system for greater long-term gain rather than focusing on short-term pain of greater competition.

Energy security in the Asia–Pacific and implications for the US

Clearly questions of supply and demand regarding Energy Security in the Asia–Pacific have been greatly exaggerated. Of much greater concern to the US government is how growing energy demand is shaping the diplomatic and trade outcomes of states in the Asia–Pacific region. While energy issues by themselves are unlikely to have a 'tipping point' impact on US regional alliance relationships, there are a number of ways energy issues could affect key relationships with Japan, the Republic of Korea, Australia, Thailand, the Philippines, and Singapore. First, energy issues involving US Asian allies such as Japan and the Republic of Korea already have had the effect of creating diplomatic dissonance. Japan's development of energy resources in Iran has been a particular source of consternation in Washington. There also has been, for a long time, grumbling about Tokyo's and others 'free-riding' on the US efforts to maintain access to Middle East oil. The 1991 Gulf War was an important episode that highlighted this difference. Dissonance in the US–Republic of Korea alliance derives not from energy issues per se, but from differences over the provision of civilian nuclear energy for North Korea. On the other hand, the US has responded positively to the Republic of Korea's offer of assistance with conventional energy for North Korea. In the case of the US–Australia and US–Japan relationships, the US decision to pursue nuclear energy cooperation with India has been met with reservations.

Second, energy concerns on the part of US alliance partners could actually help shore up US alliance relationships in the region. For example, Japan, worried about the consequences of China's growing energy demands, could use the protection of its energy supplies as an important justification not only to strengthen its alliance relationship with Washington, but also to drive its own moves towards foreign and security policy 'normalcy'. And China's military modernization, partly flowing from a desire to protect access to energy supplies, could have the effect of making several Asian states keen to buttress their ties to Washington.

Energy issues are also having an important impact on relations between the US and non-allies such as China, India and Russia. In China's case the story is, as might be expected, complex. The rapid rise of China's demand for energy and its international activities related to ensuring energy access have been the single most important driver of American attention to energy security issues in the Asia–Pacific. The now-abandoned possibility that a Chinese oil company would buy UNOCAL led to a major controversy in the US, including the holding of congressional hearings. There are also long-term concerns that China's seemingly insatiable demand for energy will contribute to naval modernization in an effort to protect sea-lanes from the Middle East. Some fears have also been expressed that China could seek power projection capabilities in order to protect its growing assets and investments in energy interests abroad. Another concern is that the People's Republic of China could be tempted to push its disputed claims to territories in order to gain access to possible energy resources. And there is concern in the US that China's heavily subsidized state oil and energy companies could undermine US and other western private energy companies in the bidding for resources abroad. Finally, there is a more general basket of concerns about how Chinese external behaviour driven by energy security concerns could complicate American foreign and security interests. For example, Chinese dealings with Sudan, Iran, Venezuela and Myanmar are examples where Chinese relations run counter to American policy and interests.

However, there are positive developments that could result from China's growing energy needs. First, China can be expected to have an increasing interest in the smooth functioning of international energy markets. No amount of Chinese efforts to 'lock-in' oil supplies will isolate China from world oil prices. Second, the need for energy could push China to pursue joint exploration and cooperative partnerships on pipelines and other mechanisms to ensure supplies. Third, energy security concerns may increase China's incentives to support nuclear non-proliferation and export control regimes, as well as maritime security and stability in key oil-producing regions. Concerns about China's behaviour regarding energy have provoked considerable alarm, but also a search for ways in which to help reconcile the meeting of China's needs with US interests. The presence of China (and India) as observers at the G8 summit in Gleneagles in 2005 illustrated the desire to bring in these major players into discussions about energy, the environment and other global issues on which they will have a significant impact. The start of a US–China Energy Policy Dialogue could lay the basis for managing US–China relations on energy issues. There is already some consideration of including China (and India) in the International Energy Agency as a way of alleviating many of the concerns of managing China's surging demand (Chanlett–Avery, 2005: 21). The possibility of US civilian nuclear cooperation with China remains on the table.

In US–India relations, energy issues have become far more salient as the overall relationship has improved and as India's demand for and activities relating to energy have expanded. In May 2005 Washington and New Delhi launched a new energy dialogue (India–US Dialogue Statement, 2005). There are three major issues on the agenda. First, the possibility of a pipeline from Iran through Pakistan to India has created dissonance between the two countries. Washington has reservations regarding the proposed arrangement. Second, and more dramatically, the dialogue raised the possibility of the provision of civilian nuclear energy to India. The February 2006 deal signed by President Bush and Prime Minister Singh is highly significant, and has attracted considerable controversy within the US and will be the subject of greater congressional scrutiny in the months ahead. Third, at the same time, efforts are underway, as recommended by the National Energy Plan of 2001, to improve India's oil and gas production and the investment climate for American companies engaged in the energy business in India. The net impact on US–India relations is difficult to say, but there are enough problems with each of these issues to inject some caution.

The impact on US–Russia relations of energy security issues in the Asia–Pacific is unclear. Russia's Far East, which possesses considerable energy reserves and proximity to North-east Asia, is of some concern to US policymakers. The primary reason is that China and Japan appear to be competing for access to those resources. On the one hand, some concern is present in Washington that Russia's energy resources, particularly natural gas, will provide Moscow with a means for greater influence in the Asia–Pacific. According to one US government report, 'If Russia continues to attract commercial and political overtures to gain access to its resources, Moscow stands to gain considerably more power in international affairs' (Chanlett–Avery, 2005). However, as with Russian arms sales in the region, it is difficult to see how and what gains, other than financial, Russia could make. A second concern is that Russia–China relations based on energy cooperation could develop into something like an axis. Again this is a very complex possibility that requires a number of factors to coincide and thus must be put into perspective.

Energy issues are increasingly important to bilateral relationships within the Asia–Pacific – most notably China–India, China–Japan, Russia–China, Russia–Japan – although unlikely to fundamentally alter the basic logics of these relationships. The impact on each relationship will of course be different, but there seems to be little possibility that any of these relationships will be altered substantially because of the energy issue itself. For example, improved Russia–Japan cooperation on energy pipelines is unlikely to result in a resolution of their territorial dispute. In the China–Japan case, energy competition and disagreement is only part of a larger package of even more fundamental problems. India and China are also looking at possible areas of cooperation (Giridhardas, 2005: 3), but here again it is difficult to imagine that such cooperation will lead to a resolution of more fundamental differences.

The impact of energy requirements on the foreign and defence policies of major countries such as China, Japan and India is also a matter of concern. Overall, one issue that causes some concern within the US is the degree of Asian dependence on the Middle East for its energy resources. As Asia's dependence on the Middle East increases, some worry that there will be more independent efforts to develop significant strategic ties between East and South Asia and West Asia. The initiation of the Asia–Middle East dialogue, among more discrete bilateral developments (Saudi Arabia–China oil deals), is the kind of development that concerns some American observers. On the other side, as Asia's dependency on the Middle East increases, it is possible that Asian countries will feel a greater interest in working with the US to provide stability there. Former Director General of Japan's Defence Agency Shigeru Ishiba, for example, recently made comments to that effect (cited in Chanlett–Avery, 2005). And even China avoided siding with Iraq before the US invasion of 2003 and reportedly has reduced its weapons sales to Iran. This is not to suggest that Asian countries and the US will see eye-to-eye on all Middle Eastern issues. On the Israeli–Palestinian issue, for example, there will continue to be strong criticism of the US and countries ranging from allies such as Japan and South Korea to friends such as India, and others such as China will maintain relations with Iran in order to ensure access to that country's energy resources.

The major US policy challenge

As the energy picture stands in the Asia–Pacific, the challenge for US policy has less to do with energy competition amongst regional states, which pipelines will be built from where to where, or even the prospect of some kind of resource wars. The ongoing challenge for the US is to reconcile a range of foreign policy and security interests that include the energy issue.

India

First, one of the Bush Administration's major foreign policy goals is to improve relations with India. Much has been done on this front – including an important agreement on civilian nuclear energy cooperation. This agreement arises from an understanding of India's energy supply problems and the effect that this could have on the future prosperity of its economy. The US sees civilian nuclear cooperation as a means to ease the pressure on domestic energy supplies and has overlooked India's refusal to sign the Nuclear Non-Proliferation Treaty by highlighting its responsible history of nuclear research. However, nuclear technology is not the silver bullet for India. Around 30 per cent of India's energy requirements are met by oil, with 69 per cent of that imported (EIA, 2005d: 30). The continued growth of the Indian economy will still require a secure source of oil and gas. Many in India see a pipeline from Iran through Pakistan as the best alternative for these supplies, even with a continued frosty relationship with Pakistan and internal instability in Iran. However for the US an Iran–Pakistan–Indian pipeline raises a range of geopolitical problems, especially considering the increasing belligerent attitude of a nuclear weapon aspirant, radical, and hostile Iranian government. The commitment of the Indian Prime Minister Manmohan Singh to the pipeline seemed to falter not long after his visit to Washington in July 2005, where he stated,

> I am realistic enough to realise that there are many risks, because considering all the uncertainties of the situation there in Iran. ... But we are in a state

of preliminary negotiations, and the background of this is we desperately need the supply of gas that Iran has.

(Interview: Indian Prime Minister Singh, 2005).

The US understands the importance of Iranian gas to the Indian economy, but clearly India's energy discussions with Iran are seen as a hindrance to US and increasingly international efforts to isolate the regime.

The complexity of this issue has been highlighted by recent attempts to bring Iran before the United Nations Security Council for violating IAEA directions. Early in 2006 US Ambassador to India David Mulford made a clumsy attempt at defining India's position in this regard. He stated,

> We have made it known to them (India) that we would very much like India's support because India has arrived on the world stage and is very, very important player in the world ... And if it opposes Iran having nuclear weapons, we think they should record it in the vote.
>
> *(US warns India over Iran stance, 2006).*

This point is valid and is in line with much of what has previously been said by the State Department. However, Mulford continued, linking India's vote against Iran with recent nuclear cooperation between the US and India. He pointed out that India's failure to oppose Iran would be 'devastating' to hopes of Congress approving the deal: 'I think the Congress will simply stop considering the matter. I think the initiative will die in the Congress – not because the US administration would want it to' (*US warns India over Iran stance*, 2006). Clarifying the threat, he stated that, '[t]his should be part of the calculations India will have to keep in mind' (*US warns India over Iran stance*, 2006). The Indian Foreign Minister admonished Ambassador Mulford's statement, asserting that 'civilian nuclear cooperation ... stands on its own merit' and that India would make a decision based on its own national interest (*US warns India over Iran stance*, 2006).

This episode highlights the difficulties that are faced in the US–India relationship. India is an important part of the US's future strategic calculations and Washington is keen to avoid being seen to be pressuring India. However, Washington also regards it as dangerous for a state such as India to work against international efforts to isolate a dangerous and nuclear aspirant Iranian regime. American, Indian, and Iranian relations are becoming more complex as the pressures of energy security force India to make difficult decisions. The issues of an Iran–India pipeline and Tehran's nuclear ambitions are only the tip of the political iceberg. What if US–Indian civilian nuclear cooperation does not live up to the high hopes that it is encouraging? While Mulford was wrong to link nuclear cooperation and Iran there is no doubt that he was probably pretty close to the truth. However, by suggesting that the agreement would be undermined by supporting Iran, opposing Iran does not suggest that it is a done deal. There are numerous ways this deal could fall through. Further, how will greater cooperation with India affect the many alliances throughout the Asia–Pacific that may be looking cautiously at India's growing economy and greater international profile?

The Koreas

The US seems to have been more successful in isolating the regime of North Korea over its nuclear ambitions than the Iranian regime. Undoubtedly this is due to Iran's abundance of energy products and North Korea's lack of them. However, energy security is at the heart of the problem as North Korea is critically lacking in energy and the credibility to receive energy aid from the US. In the past the US has stated that it is not willing to participate in energy aid to North Korea offered by other members of the six nation talks – the Republic of Korea, Japan, China, and Russia – until there is a complete, verifiable and irreversible dismantling of the North's nuclear program. This reflects the US position that it 'can not accept another partial solution that does not deal with the entirety of the problem, allowing North Korea to threaten others continually with a revival of its nuclear program' (Kelly, 2004). As it stands, the US will not support aid programs that allow the North to get off the hook. The US opposes civilian nuclear cooperation with North Korea that the Republic of Korea (and Japan) has funded and sees as a solution. The US would also oppose other ideas of a pipeline through North Korea from Central Asia that have been put forward by those sympathetic to South Korea's 'Peace and Prosperity' policy as this would supply the North with transit payments without a commitment to dismantle its nuclear program (cited in Chanlett–Avery, 2005). Of course this is not a narrow question of energy, but a larger disagreement about the nature and types of engagement with North Korea. The US is keen to engage North Korea in a multilateral fashion whilst maintaining that it will not negotiate until the nuclear issue is resolved. This approach suffers from the dual problem of undermining the solutions proposed by the other participants in the six-party talks while watering down the pressure on Pyongyang.

China and Japan

While the China–Taiwan issue remains the prime flash point in the Asia–Pacific, the rising tensions between China and Japan over securing energy supplies must also rate as a major concern that seems to have little chance of reaching an amicable solution. As the Chinese economy continues to boom and its energy needs continue to grow, China and Japan are increasingly competing for the same energy resources. Recent negotiations between Russia, China and Japan about the best route for Russian gas through Siberia were a complete farce, highlighting the level of distrust amongst all involved. Japan and China have also expended a great deal of energy to secure Middle Eastern oil.

Most worrying to the US is how they have both looked to Iranian energy at a time when the Iranian regime is on the international radar over its continued nuclear research. Further, Japanese and Chinese attempts at securing a local energy supply will no doubt fuel the ongoing dispute over demarcation of the exclusive economic zones and the Senkaku/Diaoyu Islands on the edge of the East China Sea. In the case of Japan, how far does the US go to side with its ally in its disputes with China over access to potential energy resources in the East China Sea? What effect will this have on the US–Sino relationship? As Tokyo requires more assurance about its energy supplies in the face of Chinese competition, will US efforts to coax a more 'normal' Japanese security role gain more traction with elements of the Japanese government? Sino–Japanese relations are based significantly on other issues beside energy; however, it is clear to see that energy security is influencing the continued competition and concerns of these states. And finally, what are the implications of US policies and decisions in the Asia–Pacific context for its relations elsewhere? In essence, the US is more concerned about how it approaches the energy issue in the Asia–Pacific as part of an overall strategic approach to the region and its relationships there.

Conclusions

The United States' interest in energy issues in the Asia–Pacific is driven by a general concern with the region's security dynamics rather than a specific concern with energy supplies from the region. Energy issues are an important element, but not a major determinant of the Asia–Pacific security environment. They could either facilitate cooperation, accommodation and reconciliation, or competition, conflict, and estrangement among the region's powers. The environmental impacts of the emerging energy situation in the Asia–Pacific may emerge as increasingly important drivers of US policy. Energy issues have an important role in at least two of the potential regional flashpoints, India–Pakistan and the Korean Peninsula. But energy issues by themselves will not shape or determine the outcome of these flashpoints. Energy issues will continue to present challenges to US regional alliances, but again, the challenges of the alliances stem from more important factors too.

Energy issues will be very important to US relations with China and India. With both countries there will be a mix of cooperation and dissonance that is only now beginning to take shape. Energy issues will be an increasingly important feature of intra-Asian bilateral relationships – with both positive and negative features. There are already visible elements of competition. However, there are also emerging discussions about cooperation. In other words, the handling of energy issues will likely reflect the overall state of various Asian bilateral relationships rather than fundamentally shaping them.

The trajectories of the foreign and security policies of key Asia–Pacific countries will also be shaped by energy concerns on their part. But barring any unexpected, sudden and dramatic developments, these trajectories will not veer off course. It is clear that China, Japan and even India are now, for a variety of reasons, becoming more activist in their foreign and security policies. Energy issues will likely contribute to these efforts rather than determine them. The effect on regionalism and regional multilateral cooperation is difficult to gauge at this juncture. The overall picture for regionalism and regional multilateralism seems quite troubled. In this context, it is difficult to see that energy cooperation could boost the efforts currently underway.

Notes

* The views expressed in this chapter are entirely those of the author and do not represent the views of APCSS, the US Department of Defense, US Pacific Command, or any other US government department or agency. The author gratefully acknowledges the assistance of Will Alker in the preparation of this chapter.

3 Japan's energy security policy in an era of emerging competition in the Asia–Pacific

Purnendra Jain

For resource-poor but highly industrialized and economically developed Japan, energy security is not a new issue. Indeed, part of the reason that Japan colonised Asian nations and went to war was due to perceived threats to its energy supplies. As Japan began to grow economically in the post-war period, so too did its requirements for energy, and it secured them through diplomatically crafted partnerships, even with former enemies such as Australia. During the two oil crises of the 1970s, Japan's dependence on external energy supplies, especially oil, became more apparent than ever. Japan's GDP dipped and commodity prices soared; Japan and its people plunged into economic hardship. In response, Japan improved relations with the Arab world to ensure the flow of oil and also took a number of domestic policy measures: reducing its energy consumption, passing a stockpiling law, and developing new technologies for alternative sources. Energy security has since become a central long-term foreign policy concern. During most of the 1980s and 1990s all went rather well. Even at the time of the 1990–1991 Gulf War, no serious supply shortages or major price hikes occurred as oil stockpiles were greater than even recommended by the International Energy Agency (IEA).[1]

However, energy security has once again taken centre stage in world politics. In particular, the post-9/11 war in Iraq has pushed oil prices to record levels thereby creating an international environment where securing energy and its safe passage are inescapable issues. Furthermore, a rising China and an emerging India have complicated the issue. Their huge appetite for energy and continuing demands from major world powers like the United States makes energy security policy extremely vexed. The global economy is at a crossroads as far as the political economy of oil and energy is concerned.

In this fast-changing political landscape, Japan's energy security has become ever more vulnerable. Many factors are at play. The most obvious is Japan's poor natural endowment.[2] The Paris-based OECD-linked International Energy Agency admits in its 2003 report, 'Energy security issues are more critical in Japan than in most IEA countries owing to its isolated location and limited domestic energy resources.' (IEA, 2003a: 7). Secondly during the Cold War, Japan was the major consumer of energy in Asia and faced very little regional competition for resources. Things have changed greatly since. The breakdown of the bipolar world system and the emergence of new issues including terrorism in world politics mean that Japan now faces many complexities.

The increasing competition for energy from China and India is one of the new complexities.[3] Japan has poor diplomatic relations with both China and India, although for different reasons. Economically, the relationship with China is robust and growing, but there are many unresolved political and diplomatic issues that stand in the way of an overall co-operative environment. There is little political trust on either side. Japan has also had only poor levels of diplomatic interaction with India despite India's keen desire to develop economic and other ties with Japan. There were few occasions, either bilaterally or multilaterally, for Japan and India to interact closely and share either political or economic interests.

Even relations with Russia and its extensive energy assets are complicated. Unresolved territorial issues and generally troubled relations with Russia stand in the way of Japan securing access to these vital resources. Finally, countries like China and India often see Japan as a political surrogate of the US through its alliance relationship. In their eyes and in many others, Japan is not seen as an independent player in the region or indeed in world politics.

What choices and policy options does Japan have for securing and safeguarding its energy requirements? This chapter argues that Japan's best options are to pursue multi-layered strategies. This should involve engaging regional competitors and other international players as much as possible and working towards co-operative energy frameworks through both bilateral and multilateral processes. Bilaterally, Japan will need to work towards improving relations with Russia and China in particular. Multilateral and co-operative frameworks could include integrating IEA members' activities with other major Asian players who are outside of the IEA process. Establishing regional forums for co-operation on energy security, both for securing and sharing resources, and their safe passage, is essential. Additionally, the joint development of resources, sharing of technology for efficient use of energy and building of trust are of equal importance. Japan will need to exercise caution and diplomacy, in particular giving due consideration to its existing relations with the United States. Along with these diplomatic juggling acts, Japan will need to give serious thought to readjusting its own energy policy, especially as it is now committed to reduce emissions through the Kyoto Protocol on Climate Change and to the Asia–Pacific Partnership on Clean Development and Climate (APPCDC).[4] Looking for alternative and cleaner sources of energy pushes Japan into the realm of greater reliance on nuclear energy and other forms of renewable sources of energy. Here too there are difficulties, especially in relation to nuclear energy, that are not insubstantial. Overall, Japan's options are limited and the diplomatic and policy challenges ahead are considerable.

The geo-strategic environment

Since the end of the Cold War the geo-strategic environment in the region has changed significantly. The US remains now as the sole superpower and the nations of Western Europe have consolidated themselves through their integration into the EU. This shift has resolved most of Europe's political tensions from the Cold War. In Asia, on the other hand, Cold War-type tensions continue on the Korean peninsula, across the Taiwan straits and between India and Pakistan. Economically, Japan has become less influential since the mid-1990s as other states have risen in prominence, while the Asian financial crisis of 1997 left ASEAN significantly weaker. Compounding these issues, the rise of China as an economic giant and political power has introduced a new regional geo-strategic dynamic. China has become a major concern for many key players, including the United States. Recently, the Japan–US alliance has intensified, partly as a result of China's growing influence. Added to this is the emerging influence of India regionally and globally, both in economic and politico-strategic terms. The Bush administration has now accorded India a special place in US strategic thinking, even to the extent of recognising India's nuclear status. Overall, China and India have added a new dynamic to regional geo-politics and world politics. The break-up of the Soviet Union, the emergence of new states in Central Asia and the global war on terror (GWOT) have added still further complexities. All of these developments have serious implications for energy security issues. Central Asia has again become a major theatre for competition among the dominant world actors, seeking influence for geo-strategic and geo-economic reasons: the cockpit revisited.

Even before the GWOT, one eminent scholar specialising in East Asian affairs argued that economic growth in Asia would lead to energy shortages, which in turn would lead to geo-strategic insecurity and ultimately arms build-up; he dubbed this the deadly quadrangle (Calder, 1997a: 5).[5] Although Kent Calder presented a rather pessimistic picture of the region and suggested that the US could play the role of modifier, other analysts, such as Robert Manning have presented a more sanguine picture while also recognizing rising competition and the potential for conflict (Jaffe, 2001a).

Resource-starved Japan

Japan does not have a single major developed oil field, and therefore depends on outside suppliers for almost all of its oil. Most imports are from the Persian Gulf (Kojima, 2005). The quadrupling of oil prices in 1973 and a further doubling in 1979 had devastating effects on the world economy, but among the industrialized nations Japan suffered the most precisely because of its resource-poor status. In the 1980s, Japan was a net importer of about 90 per cent of its total energy requirements. It since has cut consumption and developed every sort of alternative energy – solar power, thermal energy and harnessing of the tides. Yet, in the 1990s, Japan still imported about 85 per cent of its energy requirements (Kojima, 2005).[6]

Japan today is the fourth largest energy consumer in the world, after the US, Russia and China. The important differentiating factor is that the latter three are also major energy producers. An archipelago, Japan is unable to build a solid energy network by its mountainous topography, seismic instability and lack of accessible energy sources like gas and oil. Hence, oil accounts for 50 per cent of Japan's energy supply, almost all is imported and 90 percent is sourced from the Middle East. In the late 1980s, Japan had reduced its dependence on Middle Eastern oil to less than 70 per cent. But the increase in dependency on this region since the early 1990s can be attributed to the development of East Asian nations such as China and Indonesia (Kojima, 2005). These countries, that were formally exporting oil to Japan, have themselves become net oil importers due to the increases in domestic consumption that have occurred with their rapid modernisation.

Japan's high dependence on oil imports is expected to continue and this dependence on a single source is a major concern, particularly when that source is the Middle East. This region has a history of political turmoil and the current war in Iraq has made the region even more instable politically. To maximise its access the Japanese government has been engaging countries in the region through human exchange and co-operation in oil-related fields, such as in the development of high-precision refining technologies. Despite this the situation remains volatile.

Despite the setbacks, Japan has had some success in reducing its dependence on oil. This has been achieved through energy diversification involving natural gas and nuclear power. Diversification and conservation has resulted in Japan's dependence on oil being reduced to approximately 50 per cent of its total energy requirements. However, overall energy self-sufficiency remains low at around 4 per cent. The self-sufficiency rate rises to 20 per cent if nuclear power capabilities are included in this assessment. (Agency for Natural Resources and Energy, hereafter ARNE, 2005: 5).[7] Half of Japan's energy consumption is expended on electricity generation. Of this electricity, more than 50 per cent is supplied through nuclear power (25 per cent) and natural gas (27 per cent).

To maintain electricity generation capacity and industrial outputs, Japan has to import 97 per cent of its natural gas requirements from overseas, and it appears likely that this dependence will increase.[8] Supply sources for gas are comparatively diversified, with most of it coming from South-East Asia and Australia. In Europe and the US gas can be transported by pipelines due to the close proximity of the oil fields. However, transporting gas to Japan involves complex procedures. The gas is cooled at the fields until it reaches a temperature of −162 degrees Celsius. This process liquefies the natural gas (LNG) and reduces the volume of the gas by a ratio of 600: 1. Thereafter it is transported to Japan in tankers lined with thermal insulation. After arrival it is reconverted into natural gas, and piped to power stations and households. As a result the processing needs, gas is expensive in Japan. As an additional complication, gas supply security is increasingly uncertain. The disruption in gas supply from the Indonesian Arun LNG plants in 2001, due to a separatist conflict in Aceh, is a typical example.

Coal provides 20 per cent of the nation's energy, and for this too, Japan is almost 100 per cent dependent on imports. Coal is likely to last much longer than other fossil fuels; it is estimated that there are about 192 years of reserves. Japan imports most of its coal from politically stable countries like Australia and Canada. One important drawback is that coal produces more greenhouse gas than any other fossil fuels, although clean coal technologies that reduce the environmental impact are being developed. These technologies can be expected to significantly increase the use of coal as a clean energy resource.

Japan has been actively increasing its self-sufficiency by developing new techniques to gather energy, such as through solar or wind-generated technology. However, the scope for development is limited by high production costs and the necessity for particular climatic conditions, such as the amounts of sunlight or wind. As a result, these alternatives are regarded as unstable sources. Despite this Japan is the world's largest generator of solar energy. Wind generation has also grown threefold during the 3-year period from 1999 to 2002. At present these new energy sources only account for about 1 per cent of the primary energy supply, but the government's goal is to increase this to around 3 per cent by 2010 (ARNE, 2005: 14).

Nuclear power is another viable source of energy for Japan. The sources of materials required to generate nuclear power are available from politically stable countries. Nuclear power also has environmental advantages, particularly in relation to greenhouse gas emission. There is no CO_2 produced during the power generation process, so the impact on global warming is almost nil. However, reserves of uranium and plutonium are limited although they are expected to last longer than oil. In addition, more than 90 per cent of spent fuel from nuclear plants can be recovered and reused as fuel. Perhaps more pressingly, safety is a major concern. Further, strong political sensitivities in relation to nuclear capabilities in the region cannot be disregarded.

Policy and strategies

In the aftermath of the oil shocks of the 1970s, energy security became the most important item in the 1980 Report on Japan's Comprehensive National Security (Chapman *et al.*, 1982: 189). While Japan was able to cope with the short-term crises, the report noted long-term policy plans, especially in recognising that the demand for energy would increase from a range of countries such as the newly industrialised countries and the less industrial countries. China and India were not great concerns at the time. Oil imports were expected to grow in the 1980s and 1990s, raising Japan's import level to that of the US, thus making Japan more dependent on overseas suppliers.

The 1973 oil shock made Japan rethink its energy policy. This led to a shift in the method of power generation towards nuclear power, and the use of gas and coal, as well as an overall reduction in oil consumption. Japanese Ministry of International Trade and Industry (MITI) 1978[9] projections emphasised an increase in coal consumption, LNG, nuclear power and renewable energy sources, including solar, wind and biomass wave technologies.

Additionally, the report recognises the importance of larger emergency oil stockpiles. Before the 1970s, Japan held only 36 days worth of stockpiles. In the late 1970s IEA recommendations were that states should hold 100 days worth of stockpiles (Chapman *et al.*, 1982: 193). The report also focussed on the diversification of both fuel suppliers and sources. Attempts to diversify suppliers resulted in a decline in reliance on OAPEC from 82 per cent in 1972 to 72 per cent by 1980. Over the same period, imports from Iraq and Saudi Arabia increased by 10 and 9 per cent. Imports from the Pacific Basin, Africa and elsewhere outside the Middle East increased by 10 per cent. In the 1980s further diversification targeted Mexico and other Latin countries experiencing difficulties in meeting outstanding debts and which therefore welcomed new income sources (Chapman *et al.*, 1982: 199). However, Japanese dependence on the Middle East has since risen to close to 90 per cent of its oil requirements.

Japan's current energy policy objectives are summarised as the '3Es': energy security, economic development and environmental sustainability. These are consistent with the IEA goals (IEA, 2003: 22). Japan's basic energy policy released by the Agency for Natural Resources and Energy in 2005 (ANRE, 2005) identifies five principles necessary to maintain energy security. First, Japan needs to develop secure stable supplies, ensure domestic reserves of oil and independently develop oil fields. Second, it needs to implement energy conservation programs. Third, Japan needs to develop and introduce diverse sources of energy. Fourth, in pursuit of the prevention of global warming, it needs to make greater use of nuclear power and renewable energy (no CO_2) and natural gas (little CO_2). Fifth, Japan needs to drive further energy sector reform in response to globalisation, making sure that all players are driven by market rules and principles.

The need to diversify sources has resulted in increasing numbers of nuclear power plants in Japan. In 2004, Japan had 53 nuclear power reactors, ranking it third in the world after the US (103) and France (57). The government's target is to increase nuclear generation by 30 per cent between 2000 and 2010, which will mean 10–13 new nuclear plants. Five more are under construction with the plan eventually to have 40 per cent of Japan's electricity supplied through nuclear power generation (Kakuchi, 2005).

Nuclear power is not without problems. The process generates dangerous amounts of plutonium, the raw material for nuclear weapons. Yet the potential benefit is that 'Japan's breeder reactor program, based on reactors that produce more fuel than they consume, could ultimately break that nation's crippling dependence on the outside world for energy by the middle of the next century' (Calder, 1997a: 6). In this process, Japan would by 2050 amass about 100 tons of plutonium – more than is contained in all the current nuclear weapons of the US and Russia combined.

Moreover, safety is an important consideration. Several terrible nuclear accidents have resulted in the public having a low confidence level in the safety of nuclear power. The Tokaimura nuclear accident in October 1999 was the world's worst since the 1986 Chernobyl explosion, and the first in Asia to reach level four on the International Nuclear Event Scale. Coping with waste is also a problem.

Burying spent fuel and reprocessing of nuclear fuel is a major issue (The Yomiuri Shinbun, 2005c). Nuclear fuel recycling is Japan's national policy, but there are problems of jurisdictional authority between the national government and the local governments over where power plants are located. The energy issue has become a focus of dispute between central and local governments. The plan to revive the controversial plutonium reprocessing plant at the remote village of Rokkasho-mura, in Japan's northern Aomori prefecture, sparked a major debate about Japan's commitment to the Nuclear Proliferation Treaty. According to some scientists, stockpiling large quantities of plutonium has no peaceful use (The Yomiuri Shinbun, 2005c).

Regionally, Japan's augmentation of its nuclear power capacity is in danger of triggering a regional race for nuclear power plants, especially from South Korea and China. The trend of increasing nuclear power sources is likely to be followed in other parts of Asia, including India. As Calder commented almost a decade ago, 'Energy demand will thus steadily propel Asia to the threshold of nuclear proliferation' (1997a: 6). Asia's fascination with nuclear power plants remains high with South Korea in the forefront. South Korea had 15 plants in the late 1990s and has plans to double the number by 2015. However, this unprecedented growth in nuclear power may bring other problems as questionable oversight and poor transparency are often cited as problems of nuclear power generation in Asia (Dawson, 1999: 14).

Although debates continue about renewable and other alternative sources, most of the energy resources are finite and oil is projected to last only until 2050. By all estimates, oil will be in demand more than any other resource and much of it is located in the politically unstable Middle East. Because of this, oil and gas prices could rise dramatically and Japan most likely would face difficulty in securing energy supplies. This predicament means Japan must seek accommodations with suppliers such as Russia.

New supply sources: the Russian pipeline project

Japan's energy policy has long recognised the possibility of securing its energy resources from the former Soviet Union and now Russia. Experts believe that Russia could be a valuable source for Japan's energy requirements, but little has happened in this direction. Calder argues that:

> Japan's geographical isolation and geopolitical position conspire to make it highly unlikely that Japan could be linked to any major pipeline system in the foreseeable future. The only real possibility would be a pipeline from Siberia to Southern Korea and Kyushu, but such a solution would have to overcome Chinese and North Korean objections and these at present seem quite insuperable, quite apart form the fact that it would merely increase dependence on the Russia, with which territorial and other disputes already exist.

> (Calder, 1997a)

Many factors influence the feasibility of sourcing Russian oil. Strategic difficulties remain from the Cold War, while political instabilities make any long-term planning difficult. There are also financial considerations. The capital cost of a pipeline venture would be very high and the risks great. But Russia's potential as a source of energy cannot be underestimated. In the 1980s it was estimated that the USSR held 35 per cent of world reserves (Chapman *et al.*, 1982: 197). However, general supply was restricted as the USSR was obliged to firstly supply customers from both Eastern and Western Europe, valuable suppliers of foreign exchange and technology, both of which were vital to sustaining the Soviet economy. Today Russia is still a potential energy source. The key considerations behind realising this potential are 'proximity, profitability and political stability', says Sakamoto (2005), former vice-minister for international affairs at MITI and former president of the Institute of Energy Economics. Russia is an obvious candidate as under President Putin it is stable, close and likely to be profitable.

Russian scientists estimate that undeveloped reserves in the Eastern Siberia oil field at Angarsk, close to both China and Japan, hold somewhere around 10 billion barrels of oil. This represents about 10 years of full oil supply for Japan. In the past Japan was reluctant to invest in Russia because of an unsettled territorial dispute but 'energy realities are forcing a change of heart in Tokyo' (Fackler, 2003: 19). Japan agreed to invest $5 billion in the Nakhodka pipeline project. China has also offered financial support. A 2003 report suggested that China was silent on the Japan-supported project but 'neither side feels it can afford to lose this one' (Fackler, 2003: 19).

Two routes for the pipeline are being considered. The first project has the pipeline running into China at Daqing. The other route heads to the Pacific Ocean, running alongside the Trans-Siberian Railway to Nakhodka. The Pacific route will be about 4,200 kilometres long, connecting Tiashet in Eastern Siberia and Perevonznaya near Nakhodka (The Yomiuri Shinbun, 2005c).

In early 2005, at a meeting with the Russian Industry and Energy Minister Viktor Khristenko, Japan's then Economy, Trade and Industry Minister Shoichi Nakagawa expressed concerns about the total output from exploration in Eastern Siberia. He worried that if the Siberian reserves could not supply Japan, South Korea and the Chinese provinces of Shanghai and Guangdong, then Japan's funding for exploration might not be financially viable. Moreover, Japan's requirements for crude oil are likely to slow down by the time of the pipelines' estimated completion date of around 2012. Nevertheless, this pipeline was the centrepiece when President Putin visited Tokyo in November 2005 and reassured his Tokyo interlocutors that Russia is committed to supply energy to Japan through the trans-Siberain oil pipeline project (Alford, 2005).

The pipeline project moves one step forward and two steps backwards because of rocky Japan–Russia relations. Japan is reluctant to fund the project and China wants the pipeline to the Chinese city of Daqing built first rather than the Japanese preferred option of linking to the Pacific coast. Moreover, China–Russia relations are improving, unlike Russia–Japan and Sino–Japanese relationships. This discord is not in Japan's interest. As one author commented:

> As Moscow becomes more disenchanted with what it perceives as Tokyo's obduracy, China could replace Japan as the main beneficiary of a trans-

Siberian oil pipeline. Taking a long-sighted approach to the situation, the establishment of a pipeline from Russia is not something for China and Japan to fight over. As countries in North East Asia share the same geographical restrictions and therefore the same interests, it would serve both parties' interests better to cooperate rather than compete, with respect to policies toward Russia or the Middle East.

(Sakamoto, 2005: 25).

It is not just the East Asian states competing for Russian resources. India is also an active suitor in the region's energy markets. India's Oil and Natural Gas Commission Videsh Ltd (OVL) holds a 20 per cent stake in the Sakahalin-1 project and is looking to invest in Sakhalin-3. India and Russia signed a memorandum of understanding (MoU) in December 2004 during President Putin's visit to New Delhi. This MoU provides a basis to jointly explore and distribute natural gas from the Caspian basin as well as building underground gas storage facilities in India (Bajpaee, 2005).

Escalating tensions or regional co-operation?

Taking a realist view, Calder considers that this energy competition in Asia will lead to strategic rivalry and represents a recipe for conflict. Robert Manning on the other hand, while acknowledging that energy demand will alter geopolitical relations, argues that the outcome could be constructive rather than divisive (Jaffe, 2001a). Naval competition exists in the sea-lanes between the Persian Gulf and North-east Asia. However, scope for co-operation between Asian nations exists, due to their joint interest in securing sea-lanes free from war and piracy, thus enabling an uninterrupted flow of energy. Political stability of key regions where energy sources originate, and safe passage of sea-lanes are in the interest of major powers in Asia and elsewhere. Manning's potential areas for co-operation include cross-border natural gas pipelines; electricity grid link-ups; joint activities in fighting maritime piracy and in establishing sea-lane security; co-operation in nuclear energy and the management of nuclear waste (Jaffe, 2001a).

However, the most significant competition appears to be taking place between Japan and China. China is the world's number two energy consumer after the US and growth there has been over 40 per cent since 2000. China is developing strategies for petroleum reserves, similar to those adopted in the US and Japan, holding 75 days of reserves in Zhejiang, Shandong and Liaoning provinces (Bajpaee, 2005).

The fact that Sino–Japanese relations are presently tense makes co-operative policy difficult even if both sides wished it. Political tensions between China–Japan have been further inflamed in recent years through China's opposition to Japan's bid for a permanent seat at the United Nations Security Council. These rising anti-Japanese feelings have been expressed most directly through attacks on Japanese facilities, upon nationals in China and through verbal attacks on

Prime Minister Koizumi's visits to Yasukuni Shrine. Notwithstanding China's 'new thinking on Japan', political distrust is at its highest point between the two nations (Gries, 2005b).

In the East China Sea, also dubbed the 'Sea of Conflict', China and Japan are competing for exploring and securing oil and gas, both sides claiming their Exclusive Economic Zone (EEZ). The competition recently turned ugly when a Chinese nuclear-powered submarine entered into Japanese waters off the Okinawa islands in November 2004. The relationship has deteriorated to the extent that Japan has now identified China as a potential security threat in its National Defense Program Outline issued in December 2004.

Japanese firms have been interested in drilling in East China Sea since the 1960s but the Japanese government would not permit them. This changed when China began drilling in 2002. Japan then reluctantly also began to think about drilling in the area. In July 2005 METI granted Teikoku Oil Company the right to explore oil in disputed waters in the East China Sea (midway between Okinawa and mainland China) near Chinese drilling platforms. The ownership of this location is contested with Beijing and Tokyo unable to agree on the dividing line between their EEZ (Negishi, 2005).

In this East China Sea dispute, both sides see each other as the culprit, and undercutting the other's interest. Many Chinese believe that Japan is the culprit in seeking to destroy China's energy security (The Yomiuri Shinbun, 2005a).[10] On the other hand, Japanese argue that China's 'voracious appetite for energy has become a destabilizing factor in the international energy market' (The Yomiuri Shinbun, 2005a). The evidence for the latter is that China is ignoring international market mechanisms for reinforcing resource procurement. This in turn is leading to tensions between the two.[11]

Further complicating the matter is that free-trade negotiations between China and South-East Asia have had a two-fold effect – increasing South-East Asian dependency on China, while effectively isolating Japan (The Yomiuri Shinbun, 2005a). Co-operation between Japan and China over energy could be viable provided both see benefit in sharing ideas, information and technology. One author mentions that Japanese technology is on an even footing with Western Europe, at least in upstream areas and this could be of immense benefit to Japan's regional partners (Sakamoto, 2005: 25).

Regional co-operation

Opportunities do exist to enhance regional and international co-operation on energy-security issues through organisations such as APEC and ASEAN+3, between bilateral contacts with Asian energy-consuming countries and with the promotion of co-operation with oil and gas-producing countries. Although Japan hosted the ASEAN+3/IEA joint workshop in December 2002, further policy dialogue is needed. Dialogue is especially important to reconcile the interests of oil-producing countries such as Indonesia, Malaysia and Brunei, with those of consumer countries.

Toichi identifies three such important areas where Japan can take measures to increase oil security. First, Japan has significant human and technological resources in relation to oil reserve systems. These resources can be shared with other Asian countries. Second, Japan should assist in the establishment of a joint reserve system, and thirdly, set up mechanisms to share oil reserves between Asian countries in times of emergency (Toichi, 2003). It is not clear what sort of emergency Toichi referred to. Obviously, because of the current dependency on the Middle East, an emergency might arise if, for example, the Strait of Malacca is closed or if a war in the region affected the supply line. Any such co-operation which resulted in China being helped, for example, would be likely to reassure the latter of Japan's benign intent.

Despite ongoing regional concerns in relation to energy security, no ASEAN countries have developed government-controlled emergency oil reserves. One option put forward by Japan and Korea is a proposal to develop a joint Asian stockpile. Oil stockpile technology is a good way of building trust and confidence in the region, and will act as a hedge if an emergent situation affects the flow of oil. Precedents have been set in the past for the successful joint management of resources. In the 1950s Germany and France instigated 'The Schumann Plan', which was a program of joint management in coal and steel resources. The pooling of these resources has had a very positive effect on their relationship.

The International Energy Forum is a group of oil-producing and oil-importing countries, whose representatives meet on a regular basis to find ways and means of co-operation to ensure that the interests of both parties are secured. At the 2002 Forum in Osaka, it was proposed that a permanent secretariat be established in Riyadh in Saudi Arabia. Another idea was to connect APEC members with the Gulf Cooperation Council of the Arab States (GCC) via a forum (Toichi, 2003). An Asian version of the IEA, as suggested by Japan through its Hiranuma initiative, has the potential to be a success.[12] As suggested by Toichi, this organisation will be different from IEA as it will go beyond the developed nations and could be a good forum for the collection and sharing of energy information, sharing of technology and human resources and financial co-operation (Toichi, 2003).

Safe passage and sea-lane security

It is in the interest of Japan, China and India and indeed all other major consumers to ensure that shipping lanes remain open and that relations with hinterland neighbours remain steady and co-operative. Several scholars have noted the importance of the Indian Ocean. According to Chaturvedi (forthcoming), the Indian Ocean is the only ocean not controlled by Western powers and it is important for China, Japan and India. Due to the concentration of international terrorism in this area, the Indian Ocean area is potentially volatile.

Many authors have noted the importance of the Malacca and Lombok straits as important lines of communication between the gulf and the Pacific in the past (Chapman *et al.*, 1982: 201). The six-mile wide Strait of Malacca is the main route for oil tankers. The largest tankers must steam around Java and through the

Lombok and Makassar Straits. Passage of ships freely and without interruptions is in the interest of all trading nations. Any terrorist attacks in this area could be highly damaging to world trade. Although no major terrorist attacks have occurred in the region, piracy is a constant issue.

Piracy

The number of piracy attack cases in South-East Asia increased in recent years from about 100 to more than 450 by 2002. The number of incidents in Asia is high – some 64 incidents in Indonesian waters alone – more than a quarter of the worldwide total (NIDS, 2004: 36). The large number of crimes occurring around Indonesian waters has earned the country the title of the 'pirate republic' (Eklof, 2005). Piracy includes robbing of large cargoes while at sea, money and goods while anchored in a port or even cases of robbery of ships themselves – the Japanese-owned Alondra Rainbow hijack, eventually seized by the Indian navy in 1999 is one such example. Since piracy of goods, money, cargo and ships involves many countries, the problem becomes truly multinational and multilateral co-operation is essential to combat piracy. The Malacca Strait lies in the territorial waters of both Indonesia and Malaysia and unless there is co-operation between the two, police action cannot take place. The Convention for Suppression of Unlawful Acts against the Safety of Maritime Navigation (SUA Convention, 1988) is an international treaty to deal with modern piracy but countries such as Singapore, Malaysia and Indonesia are not parties to the treaty and they doubt its effectiveness.

Japan has actively helped the region in combating piracy through the provision of training programs and equipment to the law enforcement authorities. At a 2004 Tokyo conference a regional agreement (Regional Cooperation Agreement on Combating Piracy and Armed Robbery against Ships in Asia) on piracy was signed by 16 countries (Eklöf, 2005).[13] Four countries (Japan, Laos, Cambodia and Singapore) have also signed an agreement to set up an Information Sharing Centre in Singapore to co-operate on suppressing piracy. However, some assessment suggests that piracy is not a major issue and most shipping companies do not regard it as terribly threatening (Eklöf, 2005).

Conclusion

Energy and security issues are inevitably interlinked. Japan's high dependence on external sources, especially from the Middle East and its vulnerability is not new. This has now taken a new dimension because of the rising competition for energy from the two rising Asian giants and greater degree of uncertainties in the Middle East. The major dilemma before Japan is how to deal with its energy-hungry Asian competitors with whom the possibility of co-operation is rather limited. Plus there are many other factors at play that make Japan's dilemma even more problematic. First, Japan's tight security links with the US are often a constraint. It depends on the US for its security through its alliance relationship and is regarded as serving the interests of its ally in the region. Second, Japan's institutional constraints such as the limited role that the Self Defence Forces can play have also created difficulty for Japan. It can neither satisfy its chief partner's increasing call for greater sharing of security provisions, endangering its alliance relationship. Nor can it guarantee its suspicious neighbours (especially China and South Korea) who harbour great fear of Japan becoming a military power again. Third, Japan has unresolved disputes with many of its neighbours – South Korea, China and Russia. The lack of a peace treaty between Japan and Russia has derailed many of the economic co-operation processes in the past and a smooth and unrestricted flow of energy from Russia in the absence of resolving the political tension will always be difficult, if not impossible.

Japan will need to come to terms with its history and get this out of its way. It is a constant irritant that is not going to disappear through Japan evading the issue. Additionally, Japan will need to take pro-active initiatives towards co-operative frameworks involving major regional stakeholders that are relatively sympathetic to Japan and need its technology, finance and political friendship (such as India and ASEAN members). At the same time Japan should seek to bring those with whom they have political issues into the fold. This may also mean Japan needs to think outside of the US–Japan security alliance as it did at the time of the 1970s oil crisis. It will not be in Japan's interest to take the back seat and let the other two Asian giants – India and China – hijack the energy agenda.

Some multilateral frameworks are also being considered. Japan has proposed an Asian Energy Consortium (Research and Information System for Developing Countries, 2005) to ensure security and sustainability of energy supply, management of energy demand, energy networks and so on. Proposals for an Asian Strategic Petroleum Reserve and an Asian Emergency Response System have also been raised. Current co-operative frameworks are weak and major players rely on their own arrangements as often they are in competition with each other and their strategic thinking is different. Through its Hiranuma initiative Japan has proposed an Asian version of the IEA.[14] The aim will be to collect and share energy information, share technology and human resources and financial co-operation among nations in Asia (Toichi, 2003). All these are ideal goals but as long as serious political problems remain unresolved, for example, between Japan and China and Japan Russia, such frameworks are unlikely to work effectively.

Notes

1 The IEA was established in the early 1970s after the oil crisis and it currently comprises 26 states including Japan. For its history and role, see Scott (1994).
2 South Korea and Taiwan are also poor in natural resources, however, their energy requirements are nowhere near Japan's and their need may not rise in the same way as those of India and China.
3 There are other medium and small players whose energy requirements are likely to go up and some in the region like Indonesia, Malaysia and Brunei are likely to be net importers of oil.

4 This is a new process that was put in place in mid-2005 to which the United States, Australia, India, China, Japan and South Korea have agreed to co-operate.

5 In his analysis, Calder did not foresee India as a player in the region.

6 By contrast the US imports less than 20 per cent of its energy needs.

7 Italy is the only country in G-7 that is a little more energy dependent than Japan, while countries such as the UK and Canada have surplus supplies of energy.

8 The suppliers of LNG to Japan are Indonesia, Malaysia, Australia, Qatar, Brunei, UAE, Oman, Alaska and Trinidad Tobago. See Nishimura (2005).

9 Every 3–4 years, the government publishes the Long-Term Energy Supply and Demand Outlook, the first in 1967 and the latest in 2001. This is produced by the Advisory Committee for Natural Resources and Energy of the Ministry of Economy, Trade and Industry.

10 This comment was made by an academic of Beijing University, Bai Zhi-li who specialises in Japanese politics.

11 Because of the unprecedented price rise in iron ore in 2005 driven mainly by demand from China, some companies in Japan suffered steel shortage affecting production of cars and auto parts.

12 The Hiranuma initiative was proposed at the 2003 Osaka IEF conference.

13 These include Bangladesh, Brunei, Burma, Cambodia, China, Indonesia, India, Japan, Laos, Malaysia, the Philippines, South Korea, Sri Lanka, Thailand and Vietnam.

14 The Hiranuma initiative was proposed at the 2003 Osaka IEF conference.

4 China's energy security

Xu Yi-chong

After more than two years of preparation, the central government in China created a national energy leading group in May 2005. The 13-member group, headed by Premier Wen Jiabao, consists of Vice Premier Huang Ju and Zeng Peiyan, as deputy directors of the group, ministers from the National Development and Reform Commission (NDRC), Commission of Science, Technology and Industry for National Defence, Ministry of Commerce, Finance, and Foreign Affairs. This ministerial-level group is in charge of developing national energy strategy, the development and conservation of energy resources, energy security and emergency responses, and energy co-operation with other countries. Three weeks later, in the midst of oil prices rising to $60 per barrel, China National Offshore Oil Corporation (CNOOC)'s unsolicited bid for the US oil company, UNOCAL, stirred up the international energy community and the US government. What do these developments tell us about one hotly debated issue – energy security?

Some see its rising energy demand as the primary reason for China to compete with the existing powers for global dominance the same way as Japan competed for energy and resources in the early 1930s – 'sixty-seven years ago, oil-starved Japan embarked on an aggressive expansionary policy designed to secure its growing needs, which eventually led the nation into a world war. Today, another Asian power thirsts for oil: China' (Luft, 2004; Kahn, 2005). China's rising energy demand is considered a threat to world peace more by American scholars and policy makers than those in Europe primarily because, it is argued, major powers that are also the most voracious oil consumers find it difficult to coexist while competing over scarce resources.[1] Alternatively, some commentators prefer to interpret China's energy demand as an inevitable development of its market reform while others emphasize that energy markets have become more globally integrated. They argue along the same lines that 'China's expanding energy interests need not necessarily pose a threat to the West or to its Asian neighbours – instead they can be used as an opportunity to integrate China into existing and new global and regional institutions' (Yergin & Stanislaw, 1998; Harris, 2003: 157–177; Andrews-Speed *et al.*, 2004: 13). The prospect of co-operation is even greater with the expansion of its market forces and intensification of global energy trade, technology transfer and cross-border investment. In other words, even though interdependence does not guarantee co-operation, it does provide more opportunities and incentives for co-operation than conflict.

How can we make sense of the opposing views of these developments? This chapter will examine the development of energy security as an issue in China by asking two questions: Does China face any threat to its energy security that will eventually affect regional and global stability? Is China a potential threat to international stability as the result of its increasing energy demand? These are two related questions yet they require an examination of different aspects of energy policies in China. Indeed, there are two sets of literature discussing the issue of *energy security* – one focusing on *energy* and one on *security* even though both talk about *energy security*. At its simplest, energy security means the security of adequate and reliable energy supply at a stable price. Primary energy comes from coal, oil, gas, hydro and other renewable sources. A large proportion of primary energy is converted to electricity and indeed, the more advanced the economy, the larger is the proportion of electricity of the final energy consumption. Consequently, the research focus should be not only on securing a supply of primary energy, but, more importantly, securing the reliable supply of final energy consumption that depends on, for example, market reforms to improve economic efficiency, alleviating transportation bottlenecks, preventing large-area power outages, such as the one in New York and Ontario in 2003 or Moscow in May 2005, protecting power plants and power grids from terrorist attacks, and energy conservation (see for example, Buchan, 2002: 105–115; Austin, 2005; McNeal, 2005). These concerns, however, are often seen as domestic politics, with little implication for international politics.

The second set of literature on energy security is supply-oriented, state-centred, oil- and gas-focused and confrontational. Supply interruptions and oil price shocks remain the main concerns for many analysts and policy makers. Focusing on the supply-side, primarily of oil and gas, and being state-centred allows the analysis to equate security with self-sufficiency – that is, when a country starts importing a large amount of oil and gas, it becomes vulnerable to potential energy sanctions and it will likely engage in political and military competition with the existing oil majors, particularly the United States (for example, see Downs (2000)). Interests of multinational oil and gas companies are often equated with those of their home countries: what they do is taken as what their governments want, and the pursuit of economic interests is no different from security and political drives. People are particularly alarmed when the importers, such as China, have territorial disputes with their neighbouring states, have different political regimes, or/and their economic and military capacity has been expanding everyday (Downs, 2000; IEA, 2000; Manning, 2000a; Ogutcu, 2003; Bajpaee, 2005c).

Undoubtedly, securing an adequate supply of primary energy is an important concern for all states. It involves issues such as undisrupted access to energy resources, safe transportation of resources, relative price stability, etc. These energy concerns not only have security implications but also can become a security issue themselves. Yet, arguing that the concern for oil security drives China's diplomatic and strategic calculus seems to place the cart before the horse (Jaffe

& Lewis, 2002: 115–134). Having recognized the importance of these issues, this paper argues the overwhelming attention and efforts of Chinese policy-makers that have been on ensuring adequate and reliable domestic supply of energy that depends on the successful market reforms in the energy sector and energy conservation. Its energy policy of encouraging its oil companies to pursue access to overseas oil and gas exploration and production has been driven more by its political and diplomatic concerns, and concern for status and influence, than concerns about energy shortages They are more a part of the broader efforts of gaining international recognition as an important player and being part of the integrated world by 'going global' than engaging in competition with the energy majors (Gries, 2005a: 401–412). The question of whether China is striving to become a world power is beyond the scope of this study. It, however, emphasizes that the debate over China as a 'revisionist' rising power or as a 'status quo' rising power covers much broader issues than energy alone. In China, political and strategic considerations may drive energy policies, but not in the reverse order, at least not during the current period, given China's limited dependence on external energy resources (for the debate see Johnson, 2003; and Kupchan, 2001).

There is a difference between striving to meet domestic energy demands and securing overseas supply. Focusing on securing overseas energy supply only leaves little room for manoeuvring and co-operation because energy supply, especially oil and gas, is considered as a fixed pie and it is about the very existence and survival of a state. It is therefore a zero-sum game – 'every barrel of oil China buys in the Americas means one less barrel available for the US' (IAGS, 2005) or 'China's gain in oil and gas would be our loss in those precious commodities' (Abelson, 2005: 7). Focusing on the daunting challenges of meeting domestic energy demand will ask for an appreciation of the balancing game that Chinese policy-makers have to play – meeting increasing energy demands while maintaining both internal and external stability. Treating the recent drive of Chinese oil companies to expand their presence in world markets as part of 'China's contemporary grand strategy designed to engineer the country's rise to the status of a true great power that shapes, rather than simply responds to, the international system' (Goldstein, 2001: 836) emphasizes the integrated global market, interdependent relationship among producers and importers and, more importantly, inseparable sides of the energy security issues – energy and security, domestic and international.

By making these distinctions, we can draw opposite answers to the two questions we have raised – yes, China does face serious concerns about energy security, at least in the sense that it is facing and will face serious energy shortages in the coming years unless some effective measures are to be taken to reduce energy intensity, improve economic efficiency of the energy industry, and practise energy conservation; and no, China does not, at least not now, and will not in the near future, sacrifice its domestic stability and peaceful external environment to challenge the energy majors in order to secure its overseas energy supply. Indeed, alarmists who argue that China is building a naval capability to protect the transport networks delivering oil and gas, selling weapons in exchange for access to oil and gas assets in troubled states, and expanding territorial claims to secure its control of the resources have over-emphasized the importance of oil and gas (which accounts to about 20–23 per cent of the total energy consumption in China), overstate government capacity and the unity of views between the government and enterprises to develop long-term and well-thought-out strategies to compete with the existing energy majors, and overlooked other energy-related crises, such as insufficient coal and electricity supplies and their implications for China, the region, and the global economy.

The first section of this chapter will discuss rising energy demand as a real challenge for the Chinese government. It seeks to paint a clear statistical picture and clarify some mysteries behind the alarmists' arguments. The second section discusses the players involved and strategies developed to meet the challenges posed by rising energy demands. In the energy sector, reforms have been gradual and incremental the same way as in other sectors and the Chinese government is no more capable here than in other sectors of finding long-term strategies and carrying them through. The third section will discuss whether China is a threat to regional and global stability as the result of its increasing demand by examining the development of the oil and gas sectors. It shows that China has behaved not much differently from other oil importers – importing as much oil as it can and from as many locations as it can find. Meanwhile, with its limited reliance on overseas suppliers of oil (about 8–9 per cent of total energy supply), China's quest for energy globally, either through trade or equity acquisition, has to be examined within a broader political and diplomatic context. This being said, the perception of China's threat to regional and international stability as the result of its energy policies has to be taken seriously. The last section therefore highlights several issues that are identified as flashpoints of potential conflicts to show that energy concerns alone are unlikely to trigger diplomatic or military confrontation. This paper will advance an argument that is different from the current literature on 'global energy security', arguing that energy is only one of the many areas where China is demanding a say in shaping the global rules. Whether we like it or not, China matters and its engagement with the global energy market must be assessed in the context of its rising power and status and the associated implications for the global prosperity and global stability. Meanwhile, such an evaluation of the situation must also take into account the daunting domestic challenges the Chinese leaders face.

Increasing energy demand

Not long ago, studies on energy security focused primarily on the supply side, especially on the remaining or diminishing influence of major oil and gas producers and exporters. The main argument was that demand management is driven by environmental priorities and, according to some authors, only affected international politics through the influence of the climate change conventions and related potential curb on fossil fuel use, especially coal (Belgrave, 1985: 253–261; Haglund, 1989; Mitchell, 1996). The recent surge of interest in energy security, however, is very much about the rising energy demand, especially in China and India. One main concern is whether and how increasing energy demand in the rising powers will affect geopolitics. To analyse its security implications, we need to have a balanced assessment of China's energy demands.

China's rising energy demand as an issue of international politics has only been recent, partly because its sudden surge in oil demand in 2004 exceeded most predictions and partly because of the publicity generated by the Chinese oil and gas companies that went overseas, seeking to acquire foreign oil assets. It is also because of the security implications of the rising energy demand – 'the seeds of what could be the next world war are quietly germinating [as] China, already a net oil importer, is growing increasingly dependent on imported oil' (Luft, 2004). Within the country, energy has been a headache ever since the reform was launched. It has drawn much political attention lately because continuing economic growth has been threatened by the possible shortage of energy supplies. Chinese leaders realize that 'energy is an important strategic issue concerning China's economic growth, social stability and national security' (Wen, 2005). Concern about China's rising energy consumption is justified because even though energy consumption has been rising ever since the beginning of the reforms, in 2001 for the first time the growth rate of energy consumption exceeded that of GDP (see Figure 4.1).

China has had one of the fastest growths in energy demand in the world in the past two decades. Its total primary energy supply increased from 767 million tonnes of oil equivalent (mtoe) in 1987 to 1,221 mtoe in 2002, an almost 60 per cent increase while the total primary energy consumption doubled in the same period. Between 1980 and 2000, primary energy supply doubled, electricity

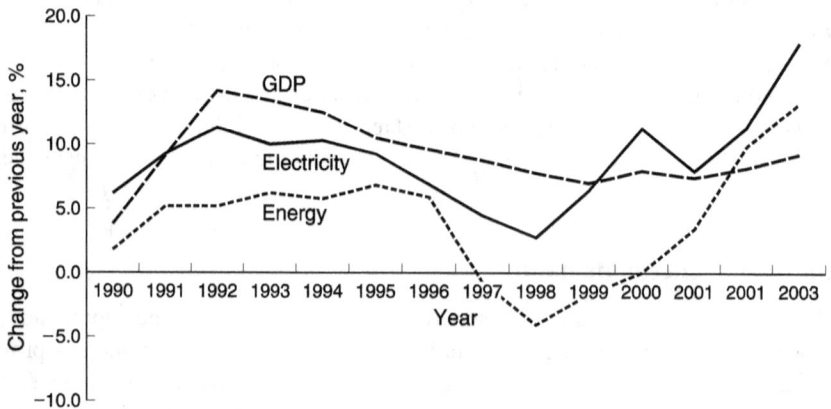

Figure 4.1 Changing relationship between energy consumption and economic growth in China.

Source: China Statistical Bureau, *China Statistics Yearbook*, various years.

production almost quadrupled, and coal production increased by almost 50 per cent. Between 1993 and 2004, oil consumption doubled as well. In 2004, energy consumption increased by 15 per cent of the year before, coal consumption 14.4 per cent and crude oil consumption went up by 16.8 per cent.

The fast-growing energy consumption was fuelled by (1) rapid general economic growth; (2) rapid industrialization – the ratio of heavy industries in the entire industrial sector grew steadily to 60.9 per cent in 2002 and 64.3 per cent in 2003; and in 2003, China accounted for 27 per cent of world steel consumption 40 per cent of world cement demand; (3) rapid urbanization – in 2003 China's urbanization rate was 40 per cent and annual energy consumption per capita in cities is about 3.5 times that of rural areas; and (4) fast growth of exports – China has become a 'world factory', manufacturing at the lower end of the international division of labour while the majority of the country's imports are high value-added products and services. Given the differential of energy consumption per unit of import and export, there has been an international transfer of energy demand.

Several points need to be highlighted to understand the challenges China is facing in terms of energy security and the impact of China's rising energy demand on the world's energy and security situation. First, until 2001, the average annual growth rate of energy consumption (about 4.6 per cent) was far behind that of GDP (between 7.4 and 9 per cent, depending on who does the calculation).[2] During the same period, the total size of the economy quadrupled while total energy production doubled. This was a 'groundbreaking change' from the experience of industrialization of other countries – that is, at the early stage of economic development, energy consumption tends to exceed that of economic growth. In the new millennium, a new round of investment-driven economic growth reversed the trend – energy consumption has grown faster than that of GDP. For every per cent increase in GDP, energy demand has grown by over 1.5 per cent since 2001. This development has alarmed China and the world because it is not clear whether 'this new energy-economy relationship in China is temporary or it indicates deeper structural change within the economy' (Logan, 2005). The trend could have significant implications for the global energy market.

Second, in the first two decades of reform, the size of the economy quadrupled while energy consumption only doubled. This can be attributed to a large extent to improved energy utilization (energy used to produce a unit of GDP) – the energy intensity (toe/US$ 1000) reduced from 2.5 in 1980 to 0.9 in 2000 and 0.8 in 2001. Meanwhile, energy intensity remains far above that of other major countries – 'energy consumption per unit of GDP stands at five times of US level and 12 times those of Japan', according to the IEA study (IEA, 2000). According to some studies, the gap is much smaller, yet it is commonly accepted that China has the potential to achieve a further 30–50 per cent reduction in energy consumption by improving its industrial energy efficiency.

Third, energy demand will continue to rise despite the potential reduction in energy intensity because energy consumption per capita in China remains far behind the world's average – in 2000, primary energy consumption per capita in

China was about 9 per cent of that in the US, 16 per cent of OECD countries and 50 per cent of the world average. In 2002, total primary energy supply per capita in China was 0.97 toe while the world's average was 1.65. Reducing unbalanced energy consumption between the rural and urban sectors and across different regions will also contribute to rising energy demand. Energy consumption in urban areas is over 60 per cent more than that in rural areas. Over time, as rural consumers catch up with their urban counterparts and regional income disparity narrows, this will place great pressure on the management of energy demand.

Fourth, coal is the principal fuel source for China, accounting for about 60 per cent primary energy production. Over 70 per cent of electricity is generated with coal. China is the largest coal-producing country in the world now and its coal reserves account for about 11 per cent of the world's total. On the one hand, the IEA estimates China's coal reserves at 114.5 billion tonnes (about 12.6 per cent of the total world coal reserves) – enough to last between 80 and 100 years (IEA, 1999: 9; BP, 2005). The issue therefore is not whether coal will remain the main source of its energy: it will; nor whether its supply will last: the medium- and long-term supply is guaranteed. Rather, China has been pushing to replace coal with other energy sources ever since it started the reform, because coal has low energy utilization efficiency, low economic benefits, especially in energy-intensive industries, and low product competitiveness. It also has serious environmental impacts by destroying arable land that is already in short supply, destroying groundwater sources that is another main shortage, and polluting the atmosphere with sulphur dioxide emissions. High dependence on coal has also had negative impacts on coal production. Driven by economic profit, coalmines, especially medium and small mines have mined only thick coal-beds and discarded thin coal-beds. The resources' recovery rates for some small mines are believed to be only 20 per cent and the problems of damage and waste have been serious (Andrews-Speed, 2004). These developments have major implications for China's energy security.

Fifth, demand for oil and gas will rise faster than coal. Oil consumption doubled between 1984 and 1995 from 1.7 to 3.4 million barrels a day (m b/d), and then almost doubled again in the next decade, to 6.4 m b/d. Since domestic oil production has remained flat, imports of oil have increased steadily since 1993. China's oil dependence rose from 6.3 per cent in 1993 to 30 per cent in 2000 and 46 per cent in 2004 (Wang *et al.*, 2005). Oil consumption will continue to rise partly because of the government's desire to change the fuel mix and partly because of the uptake of family cars (after tripling between 1995 and 2000, the number of private cars doubled between 2000 and 2003). In 2002, energy consumption for transportation and telecommunication services in China accounted for 7.5 per cent of total energy consumption, an increase from 4.6 per cent in 1990. Among OECD countries, energy demand for transport is about 33 per cent and the transport sector alone is responsible for over 60 per cent of oil consumption. Private automobile ownership increased steadily in the first couple of years of the twenty-first century, from 0.3 cars per 1,000 people in 1985 to 1 car per 1,000 people in 2003. This still lags behind many developing countries, 27 in

Thailand and far behind developed countries, 498 in Germany (International Road Federation, 1998; NBS, 1998). The speed of automobile development in China will place serious pressures on the energy supplies, especially oil.

Finally, what do these developments mean for regional and international energy security? Since rising energy demand in China has been fuelled by rapid economic expansion in the past half dozen years, especially in its export industries, China's economy has become very much part of the global economy. China by now has become the chief economic driver of Asia, leading not only the South-East Asian countries that suffered during the Asian financial crisis, but also Japan, out of their economic doldrums. Globally, world oil price increases in 2004–2005 have not had similar disastrous impacts on the world economy as the two oil crises in the 1970s and 1980s, according to many analysts, mainly because the cheap imports from China curbed inflation pressures. Meanwhile, energy shortages that could choke China's economy and lead to serious economic slowdown will likely have a much greater impact on an integrated world economy than a decade ago. Competition for access to global energy sources can trigger conflicts with existing major powers. Either way, as some international energy experts put it: 'the means by which Beijing chooses to deal with its energy security will not only affect the Chinese economy, but the global economy as well; China's energy needs have global implications today' (Dorian, 2005).

Daunting challenges for the Chinese

Is China facing an energy crisis due to its rising demand? Is China's crisis turning into a threat to regional and global stability? The alarmists have expressed concerns since China became a net oil importer in 1993:

> A second tier power whose foreign concerns (beyond nuclear issues) were mainly about defending its borders is becoming a global player with interests extending through Eurasia to the Middle East and to North and West Africa. The quest for oil has taken Beijing as far afield as Latin America. It is also affecting its attitudes toward US foreign policy.
>
> (Jaffe & Lewis, 2002: 115)

Rising energy demand and recent energy shortages have posed an even greater challenge to Chinese policy-makers. China at the moment faces three shortages: electricity, coal and oil. After a short period of electricity surplus at the end of the 1990s, 24 out of 31 provinces in China have experienced severe power shortages in the past 3 years. Power shortages have triggered shortages in coal, which generates almost 70 per cent of electricity in China, and in oil that is used to compensate for the shortage of coal. Energy shortages are not a new issue; indeed the problem was much more severe when reform started in the 1980s. What is different, however, is the ability to mobilize the necessary resources to deal with the problem. Whether and how China is able to alleviate its energy shortages will have direct impact on the economic wellbeing and political stability of the region

and the world. Two issues are at the centre of the concern – China's organizational capacity and its energy development strategy.

In contrast to the argument that concerns about oil security are increasingly
influencing China's diplomatic and strategic calculus, it is argued that the main
energy concerns in China remain domestic – that is, all the strategies and efforts
to alleviate energy shortages are targeting domestic supply and demand rather
than pursuing aggressive external policies, even though how China deals with the
issue of energy shortages will have direct impacts on the economy in most parts
of the world. This can be appreciated when the whole energy sector is taken into
consideration, not just oil and gas that together account for about 18 per cent of
total energy production in 2003 and 25 per cent of total energy consumption
(NBS, 2004). Due to financial, ecological and technological limitations, the
development of hydro, nuclear and other renewable energies will continue to play
a subordinate role in China's energy mix.

If energy resources endowment is a key factor in deciding what energy strategies a country can adopt, government capacity is the crucial factor in deciding
how the strategies can be translated into policies and programs and how they can
be carried out. The following section therefore provides a brief summary of actors
involved in the energy sector and the development of China's coal, electricity and
petroleum industries in the past 25 years.

Players

China is one of the few countries in the world that does not have a national government agency co-ordinating energy development. In most countries, including
the United States, which is known for its 'free' market system, a national government department is in charge of developing national energy strategy, protecting
energy security, and supporting and co-ordinating energy research and development. In most countries, including the US, the energy sector is dominated by
either a monopoly or an oligopoly in each sub-sector – a few companies control
up-stream and down-stream production and marketing is a common practice
everywhere in the world and vertically integrated monopoly has been the norm in
the electricity sector until very recently (see for example, Newbery, 1999).

China has not had an integrated authority co-ordinating all energy sub-sectors
and energy policies since 1992 when the Ministry of Energy was dissolved and
the agencies overseeing the three sub-sectors – coal, electricity and petroleum –
went their own ways. The coal industry is the most decentralized. The Ministry
of Coal (1992–1998) evolved into the State Administration of Coal Industry,
SACI (1998–1999). When SACI was abandoned in 1999, no one authority took
over its responsibilities – preparing strategies for the national development of the
coal industry, establishing guidelines, policies and regulations for the industry,
organizing the sale, allocation and transport of coal from state-owned mines,
improving the efficiency in production and co-ordinating relationships between
the coal industry, government and associated ministries (IEA, 1999: 11). Unlike
in India where heavy reliance on coal for energy has led to a centralized

organizational structure led by a mega state corporation and the Ministry of Coal, in China power struggles within the old ministry prevented the formation of a national coal corporation. Instead, the China Coal Industry Association, whose membership is voluntary, was created in 1999 to promote market reform, draft industry regulations concerning quality, technology and management standards, and enhance foreign co-operation. Its effectiveness has been minimal.

Furthermore, as a result of the economic reforms, the coal industry has become decentralized in terms of ownership of coalmines. Real power is in the hands of provincial or local governments that have no intention of submitting to any sort of oversight from an association. Since coal is the only source of revenue for local governments, illegal coalmines and wasteful mining have become a common practice and deaths from coal mining accidents increased steadily from 5,000 in 1980 to 7,000 in 2003. In 2002, the average death rate (person/Mt) in China was 188 times of that in the US and 15 times that in Russia. The death rates for town and village mines are double the national average (Andrews-Speed, 2004; Wang *et al.*, 2005). In addition to human costs, illegal mining has led to serious inefficient mining and depletion of key coalmines. As experts have correctly pointed out, 'if coal problems persist or even worse, Beijing could be forced to dramatically increase emphasis on nuclear and imported oil and gas options to meet long-term energy requirements, possibly raising concerns about reactor safety and Chinese reliance on Persian Gulf energy' (IEA, 1999; Dorian, 2005; EIA, 2005h). Indeed, in 2004, the China National Nuclear Corp announced that the Chinese government was planning to increase the installed capacity of nuclear power plans to 36 GW by 2020 with the adoption of new generation reactors 'featuring advanced security, high heat efficiency and a simple operation' (Lan, 2004: 28).

In contrast to the coal industry, the electricity sector was centralized until very recently. After the Ministry of Energy was dissolved, Li Peng, then the premier of the country, was determined to put the industry together and maintain its vertically integrated structure. The Ministry of Electric Power was eventually replaced by the State Power Corporation of China (SPCC) in 1997. In the first 5 years, SPCC was able to corporatise most state-owned enterprises in the sector, expand the generation capacity and extend transmission and distribution networks. Consequently, the industry was able to alleviate power shortages to the point that there was a sudden and 'mysterious' surplus at the end of the last century. In December 2002, the vertically integrated SPCC was unbundled into two grid corporations, five national generation groups and four auxiliaries. Coincidence or not, since the unbundling, China has experienced power shortages in 3 consecutive years (Lieberthal & Oksenberg, 1988; Xu, 2002). A power shortage in the Yangtze River Delta was rapidly followed by coal and oil shortages stretching across the north-west, east, south and south-west of China. To avoid blackouts, especially to meet peak hour demand, many utilities in 2004 imported a large amount of diesel and this became one of the main reasons for the rising oil prices in the world market (Eckaus, 2004).

Oil and gas account for a small portion of the total energy supply in China (14.4 per cent in 1990, 16.6 per cent in 1995, and 22.6 per cent in 2001) (IEA,

2003c: 18). The attention to it, however, often overshadows the fact that coal remains the backbone of the energy consumption in China and of the country's economy. This being said, we will have to recognise the importance of oil and gas in China's economy and its relationship with the rest of the world.

One important defining feature of China's energy policy regarding oil and gas has been and remains self-dependence. China was an oil importer in the 1950s and its reliance on the Soviets for oil exploration and production (E&P) eventually made China pay a very high price when the USSR pulled out from all the projects in the middle of construction (Woodward, 1980). Since then Chinese leaders have been so determined to ensure energy self-sufficiency that opening the oil industry to foreign investment was one of the very first reform measures initiated at the end of the 1970s. In February 1982, the Chinese government adopted the Regulation on Offshore Exploration by Foreign Cooperation to attract foreign investment to satisfy its thirst for foreign technology and equipment, which could only be financed by selling natural resources, and to revive this industry which had been in disarray since the Cultural Revolution (Lieberthal & Oksenberg, 1988: 228).[3] To achieve these objectives, the State Council in February 1982 created the China National Offshore Oil Company (later incorporated as CNOOC) as an interface organization under the Ministry of Petroleum Industry. CNOOC was to conduct exploration and production in China's offshore areas both independently and as the exclusive Chinese partner for foreign entities. GNOOC was designed as 'the agent through which China would assume an equity position in the development stage, should commercially viable quantities of oil be found off shore' (Lieberthal & Oksenberg, 1988: 125). CNOOC had full responsibility for all policy matters relating to offshore oil activities in the areas under its jurisdiction, including bidding, carrying out negotiations and ruling on any issues of contract interpretation that arose. Its four branches – the Bohai, South Yellow Sea, Nanhai East and Nanhai West – were all designed to expand China's oil exploration offshore. In 1998, the government 'allocated the management of China's LNG imports to CNOOC's portfolio' as well (Chang, 2001: 231).

When the Ministry of Petroleum was dissolved in 1996, three leading state oil companies – China National Petroleum Corporation (CNPC), CNOOC, and China National Petrochemical Corporation (Sinopec) – were promoted to the ministerial level and placed under the State Economic and Trade Commission (SETC). Along with the promotion, they were delegated the power to purchase operating rights and rental rights overseas, and to establish subsidiaries to undertake overseas oil exploration (Troush, 1999). CNPC focused on petroleum exploration and production while Sinopec focused on oil refining and distribution.

In 1998, another round of government restructuring created an oligopolistic structure in the oil and gas sector following the model of western oil and gas companies (IEA, 2000: 37). The state oil and gas companies grouped under the State Administration of Petroleum and Chemical Industries, a newly created regulatory body, were allowed to compete freely, both domestically and internationally, and across the spectrum of exploration, production, refining and marketing. With the restructuring, CNPC and Sinopec divided the country between them: CNPC, with the most domestic crude oil resources took over the control of the north and west and Sinopec, with the most developed markets and access to foreign oil took over the more developed south and east regions in China. Meanwhile, China National Star Petroleum Corporation was created to introduce competition but it was soon taken over by Sinopec in 2000 in order to build the Chinese oil and gas companies into large-scale corporations. CNOOC's position in offshore exploration and production was confirmed.

The three state oil companies are fundamentally different from the old state-owned enterprises. They are neither the puppies of the government nor have complete independence. Instead, they have their corporate interests to consider and their corporate interests are not always in line with those of the government or the CCP (see Pei, 2005). While they gained much greater autonomy in making decisions concerning their operation, they maintained close relationship with SETC and other government agencies because of the nature of Chinese politics. For example, 'the previous Chief Executive of Sinopec was appointed Chairman of the SETC' and the previous Chief of Executive of CNPC was appointed the minister of the newly created Ministry of Land and Natural Resources (MLNR) that is in charge of land use (Andrews-Speed, 2004: 176). Having political and economic clout over decision-makers does not mean these oil companies can make decisions on behalf of the government. Furthermore, if, as studies have shown that, 'the Party no longer monopolizes nationalist discourse' in China, how and why do we assume it can monopolize economic policy-making that concerns not only domestic interests (both public and private) but also the interests of overseas players? (Gries, 2005a: 402). With six different government ministries and departments having jurisdiction over oil and gas sectors with only about 50 professionals spreading these ministries and departments, the argument made by alarmists that actions taken by CNOOC or CNPC or any other state corporations were part of a long-term strategy worked by the CCP and the Chinese government with the intention to replace the US as the dominant power clearly exaggerated the capacity of the Chinese government to work out such long-term strategies and force state corporations to do what it wants them to. It also overstates the idea that the government knows what it wants to do. It has long been accepted that reform in China has been gradual, incremental and piecemeal and there has not been a grant strategy or a long-term plan (Naughton, 1995). There is no evidence to show that energy is an exception. Rather, the negotiated relationship between the government and these state corporations provides another dimension to this interpretation.

At the beginning of the new millennium, the tightness of domestic energy supplies forced the government to rethink its energy policies. The first step of a new strategy was to form an integrated central authority to co-ordinate energy supply and demand. 'China's dysfunctional energy bureaucracy' drew a lot of attention not only among Chinese policy-makers but also from international organizations, such as the IEA and the World Bank, which see it as the main contributor to energy shortages. Yet, the central government had great difficulty creating a central body that was able to address the country's overall energy requirements. Once power and authority had been decentralized, neither provincial and local governments nor state enterprises wanted to give up their control or be subject to regulation in the way market forces are used in developed market systems. This fragmented and competitive energy sector is different from what Lieberthal and Oksenberg (1998) described as 'fragmented authoritarianism'. It remains fragmented but the government is no longer able to facilitate the energy development by controlling allocation of resources. Its indirect policy instruments, such as project approval, loans, or regulations, are as ineffective as central control.

In April 2004, the government created an Energy Bureau under the NDPC as an integrated central authority responsible for studying energy development and utilization both at home and abroad, developing long-term energy strategies, making recommendations to the State Council on energy policies, and administering oil, natural gas, coal, electricity and other parts of the energy sector. At the same time, it was given full authority over the development of national oil reserves. The Bureau, however, was equipped with only 30 staff (Wang *et al.*, 2005). By contrast, the Office of Policy and International Affairs, one of the 16 divisions of the US Department of Energy (not including its operational offices and laboratories), has more than 90 staff; the 30 staff in the Chinese Energy Bureau are clearly insufficient to undertake all of their responsibilities. Meanwhile, the large-state and semi-state corporations dominating each sector have insisted on doing what they wanted to do in pursuing profits and maximizing shareholder income (see the specific examples provided by Downs, 2004: 21–41).

When the rising demand for oil drove up world oil prices and became an international issue, the central government in 2005 once again tried to create 'an authoritative institution' to achieve the combination of supply expansion and demand management. The 13-member national energy leading group was established in May 2005, with Premier Wen Jiabao as the director. It was in charge of making energy strategy concerning the development and conservation of energy resources, energy security and emergency responses and energy co-operation with foreign partners. The group would also co-ordinate and supervise existing energy authorities, provide a master plan for the development of national energy resources, and finalize and enforce energy policies. Finally, it would advise the State Council on sustainable energy policies. A 24-member executive office was established at the NDPC to help the leading group oversee the macro energy growth trends, organise research, and manage other administrative matters; just 24 people. Meanwhile, the state energy corporations, from the three oil companies to the four electricity corporations, insist the leading group is no more than a body making broad policies, whereas they are the ones who decide how they operate and where they want to invest.

Strategies

China's economic development has significant implications for national and global energy supplies. In the first two post-reform decades, in all three energy sub-sectors, the emphasis was on reforming the industry and making it more-efficient to meet domestic demand. The main foreign policy concern at the time was to ensure foreign investment and China was quite successful in attracting interest in the energy sector (IEA, 2000; Ogutcu, 2003). There was not an integrated policy directing the country's energy development. Policies adopted in the sector were disjointed, often fixed on multiple, mutually exclusive objectives, and designed to meet political ends at the expense of economic considerations. This should not be a surprise because it is widely accepted that Chinese economic reforms have been experimental, gradual, and incremental. China has not been able to follow any grand strategies and there is no reason to expect these strategies will be realized.

Increasing dependence on imported energy resources can make any government uncomfortable, even more so for the Chinese government with its bitter historical experience. An energy strategy suddenly became urgent when rising energy demand coincided with flat supply and rising demand alarmed the international community. However, efforts to develop a national energy strategy should not be over-stated. The fact that three strategies were developed in 5 years with different emphases shows that they are more indicative than working as grand plans. Moreover, while energy has been a key issue for the Chinese government and energy industries, its foreign policy concerning energy is about much broader strategic and diplomatic issues than energy alone. Beijing's desire to be recognised as a major player in international politics and to gain a seat at the international energy table has to be taken into consideration when these energy strategies are analysed. China, with a share of 12 per cent of the world's total energy consumption, is a significant player in the global energy market in its own right. It wants to be treated as a major player and demands its 'right' to share the world's energy resources.

In November 2001, the SETC worked out an energy strategy consisting of seven points:

1 Increase efficiency in coal consumption;
2 Develop domestic oil exploration and production;
3 Promote the 'going abroad' plan of state oil and gas companies;
4 Increase security in oil markets;
5 Rationalize pricing of oil markets;
6 Develop utilization of natural gas; and
7 Promote alternative energy technologies (Lewis, 2002: 13–14).

In 2003, the Energy Research Institute of NDRC and the Development Research Centre of the State Council worked out several versions of China's National Energy Strategy for the twenty-first century. The listed priorities for the Chinese energy policy included:

1 Increase investment in basic exploration of all energy resources;
2 Improve the management of energy resources;

3   Give priority to the development of hydropower, to accelerate the development of nuclear power and to develop the new and renewable energy;
4   Speed up the implementation of 'two resources and two markets' and fuel mix changes;
5   Improve energy supply security for China;
6   Improve clean-coal technologies;
7   Promote the utilization of high-quality energy (Development Research Centre of the State Council, 2003; see also Andrews-Speed, 2005: 13–17).

While most of these policy recommendations were about domestic policies and required changes in domestic energy sectors, external attention has been on number five – securing energy supplies for China (Marcois & Miller, 2005). This particular policy recommendation included four specific aspects: (a) to develop multiple import sources and import locations by increasing oil imports from Russia and Central Asia, raising the proportion of crude oil import from Africa and Latin America, and diversifying oil imports from the Middle East to different countries; (b) to build up oil reserves to avoid unexpected interruption; (c) to promote and strengthen regional and bilateral energy co-operation; and (d) to participate in the Energy Charter Treaty. Both strategies of 2001 and 2003 focused on oil and gas. This was partly because they were designed in the shadow of the war on terrorism in Afghanistan and later in Iraq. Many in China at the time argued that 'it relied too heavily on imports from the Middle East, ... the source and transportation route for those imports will be the major concern that influences China's energy security, especially when the supply area is considered unsafe' (Zhai, 2003: 19). To secure and diversify its access to overseas energy resources, the government listed 'three strategic regions' for the Chinese oil companies to target – Central Asia and Russia, the Middle East and North Africa, and South America. This part of the energy strategy has drawn the most international attention and alarmed those who are sceptical about Chinese ambitions.

After two consecutive seasons of power shortages in 2003 and 2004, two main think tanks attached to the State Council and the NDPC revised the country's energy strategy. The new strategy was quickly endorsed by the Chinese government and placed its energy priorities on:

1   Building large coal production bases;
2   Enhancing coal mining technologies and improving coal transport;
3   Expanding power projects, power grid construction and the development of key equipment in the power industry;
4   Developing new energy sources, including nuclear, wind and solar power projects; and, more importantly;
5   Energy conservation, especially calling for tight control over industries with high energy consumption (*Energy Leading Group*, 2005).

At the very first meeting of the national energy leading group, Premier Wen Jiabao called for a full understanding of the significance and urgency of energy work in China. 'Energy is an important strategic issue concerning China's economic growth, social stability and national security', said Wen (*Energy Leading Group*, 2005). Yet, the policy priorities set up targeted primarily domestic supply and demand – maximizing domestic supply and improving energy conservation, both of which require better planning and better co-ordination among different sectors of the economy and calling for a centralized 'authoritative institution' to make energy policy for the country.[4]

Some international observers see the drive for developing domestic energy supplies or restoring energy self-sufficiency as a pipedream because of China's limited reserves.[5] Others see these strategies as a bunch of conflicting views that will not get China anywhere until it can develop some consensus on what China wants in its energy development – development of indigenous oil and gas, diversification of energy sources and imported energy supplies, or encouragement of energy conservation and efficient energy use. At the moment, China seems to be moving in each of these directions – pursuing diversified, secure import sources, actively attracting foreign investments in its energy sector, investing in overseas production facilities, engaging in massive domestic development programs of which energy is one major part, trying to create strategic oil reserves and improving energy conservation. Among these, energy policy has become increasingly a subset of foreign economic policy in general. Some question whether China can achieve all these objectives because there are some inherent conflicts between several of them. Others argue that these measures are not necessarily contradictory and they can co-exist and be complementary with each other (for the debates and conflicting interests, see Downs, 2004; Austin, 2005; Constantin, 2005).

Oil and diplomacy

Oil has become a focal point of attention not only because it is a strategic commodity crucial to national economic growth and the country's strategic position but also because of the fundamental imbalance in the world's oil reserves. With two-thirds of world oil reserves located in the Gulf region, the uneven endowment of petroleum can easily be translated into political and diplomatic problems for those who have to import energy. For example, current estimates show that Asia accounts for about a quarter of world demand for oil but has only 10 per cent of supply. This resource gap is most pronounced in Northeast Asia. Japan has no significant domestic oil and gas fields and has to import 99 per cent of its oil needs. Korea and Taiwan similarly rely on imported oil. Regarding oil and gas, there are three issues involved – trade, equity investment and building national strategic energy reserves. Any moves by China to acquire oil through trade or investment are closely watched.

Trade

Since 1993, while China has increased its oil imports steadily, it has also made conscious efforts to diversify its import sources. In 1993, for example, almost all of China's crude imports came from Indonesia, Oman and Yemen. By 2004, Saudi Arabia was China's largest supplier, accounting for 14 per cent of its imports. China's imports have raised three concerns: one is the standard zero-sum argument – that is, if China takes one barrel of oil, it would mean one barrel less to other importers. The argument, however, is seldom phrased in this fashion because petroleum has a global market and everybody can come and purchase oil and gas at the right prices. Instead, the argument is often disguised by the argument concerning China's diversified locations – China has been building a special relationship with Saudi Arabia, a traditional US ally in the Middle East, in order to compete with the US for influence in the region (Lewis, 2002; Woodrow, 2002; Luft & Korin, 2004). The third argument is that China's dependence on oil import from the Middle East constrains its foreign policy options.

The counter-arguments would be: first, China has been doing exactly what all other oil importers do – diversifying and increasing its oil imports from several oil exporters. Trade is one of the most effective mechanisms helping China to get where it is now. It will continue to play the card. Second, increasing oil imports from the Middle East do not imply the reduction of imports from its traditional suppliers in Asia and the Pacific. Indeed, 'China imported more than five times as much crude from [Asia–Pacific suppliers] in 1997 as in 1990' (IEA, 2000: 51). Third, other studies have shown that despite its increased oil imports, 'China's attitude toward the Middle East has tended to resemble that of a disinterested bystander' (Andrews-Speed *et al.*, 2003: 25). This leads to the fourth counter-argument – unlike the US, China remains a developing country and it does not have the same broad foreign policy objectives as major powers do. The subject matter is beyond the scope of this paper but it is important to note that despite its rhetorical statement of anti-hegemony and pro-multilateralism, the Chinese policy-makers on many occasions have emphasized the importance of co-operating with the US. It does not hesitate to be a free rider for now and its foreign policy with its limited objectives seems to be more domestic-interest oriented than targeting any major powers or regions.

Equity purchase

Building *two markets* and *two resources* to diversify its energy sources is neither a new strategy nor an independent one from other policies. The petroleum industry was 'at the forefront of forming joint ventures with foreign firms in which the foreigners obtained equity holdings in China' (Lieberthal & Oksenberg, 1988: 169). China made its first onshore petroleum discovery in the Daqing oil field in the north-east region in 1959. This and other large onshore fields were more than sufficient for national consumption throughout the late 1970s. As demand accelerated and production of key onshore fields declined, further exploration and development of reserves became important, but initial exploration and development required foreign technology, expertise and capital.

Despite extensive exploration by Western firms, such as Arco, Exxon–Shell, and Texaco–Chevron, in the East and South China Seas, only small pockets of oil were found, many of which were not commercially exploitable. In 1993, for the first time, China joined the ranks of international net oil importers. The official policy on oil development at the time was 'to find new sources of oil production all over the country, and especially in the western part, to compensate for the likely loss of producing capacity in the eastern part' – 'stabilizing the east, developing the west' (Fu & Li, 1995: 170). Meanwhile the country reshaped its foreign policy. 'In the report of the 14th Party Congress held in 1992, China, for the first time, clearly put forward the strategic policy of actively enlarging the overseas investment and multinational operations of China's enterprises' (Zheng, 2004: 27). This report sent a green light for the three oil companies to go out, seeking oil equity on global markets.

By the mid-1990s, it had become clear that China did not have the oil reserves to satisfy its increasing demand. Given its rapid economic growth and limited oil reserves, Li Peng, then the premier of China proposed in 1997 'development in the petroleum sector should rely on *two markets* and *two resources*' – getting involved in exploration and production of oil and gas in the country and overseas (Peng, 1997). To secure strategic sources of oil imports, the Chinese government adopted the '*going out*' strategy – encouraging its oil companies to share overseas oil and gas resources. This energy strategy was adopted in line with the general reform policies at the time – to convert loss-making state-owned enterprises to a modern enterprise system through corporatisation and forming large enterprise groups, a policy adopted at the 3rd Plenary of the 14th Party Congress in 1993 (Chow, 1997; Xu, 2002: Chapter 4). The Chinese oil companies were among the first groups which were encouraged to build themselves into world-class large corporations.

Between 1993 and 2000, CNPC was the main oil company going out to secure either oil exploration and production rights and/or refinery assets. It aroused little international attention. In 1993, for example, CNPC purchased partial rights of an oil company in Alberta, Canada, and started producing the first Chinese overseas oil. In the next 2 years, CNPC secured contracts in oil production in Peru and Sudan. In 1997, CNPC, with about US$5 billion, obtained a 60.3 per cent share of an oil company to develop two oil fields in Kazakhstan and started construction of a pipeline to export oil eastwards to China. CNOOC started overseas investments as well. In 1993, it acquired a 32.58 per cent interest in a block in the Strait of Malacca by purchasing shares from Arco. It purchased an additional 6.93 per cent share in 1995 to become the largest stakeholder. Even though none of these overseas investments brought any substantial oil supply to China, by the end of the century China had become one of the largest global investors among developing countries, with its total outward foreign direct investment stock at the same level as that of South Korea, over US$25 billion (UNCTAD, 2004).

By the second half of the 1990s, building large corporations and putting them on the list of Fortune 500 had been accepted as an official policy not only to develop the Chinese economy, as large corporate groups did in Japan and South Korea, but also to put Chinese corporations on the map as part of the efforts to build China as a world power. As the chairman of CNPC put it: 'overseas operations make up 60–70 percent of the total business of global giant like ExxonMobil. Over the long term, we are working toward that goal' (Zhai, 2003: 20).

In early 2001, Chinese premier Zhu Rongji again called for the Chinese enterprises to implement a 'going-out' strategy and invest in the world. '*Going-out*' was part of the broader policy of global engagement. Overseas oil and gas equity purchases by the Chinese oil companies have only been part of the political, diplomatic and economic game for the Chinese government. To the Chinese, the country could not develop without opening up to the outside world; opening up means both allowing foreign companies to come in and Chinese corporations going out to invest. 'Going out' to develop and utilize foreign resources, and diversifying its own oil imports, need co-operation; co-operation requires a mutually beneficial situation for all involved.[6] Based on this principle, China's oil companies accelerated their hunt for overseas oil assets as part of the country's larger 'going out' strategy. Growing exchange holdings made these purchases possible. The largest deal ever made by CNOOC was its US$585 million purchase of the equity of Spanish interests-controlled Repsol–YPE's five oil fields in Indonesia. It included assets in five major oil blocks – South East Sumatra (65.34 per cent), Offshore NW Java (36.72 per cent), West Madura (25 per cent), Poleng TAC (50 per cent) and Blora (16.7 per cent). In total the deal brings China about 40 million barrels of oil a year. In 2001, CNOOC acquired 12.5 per cent interest in BP's Tangguh LNG project in Indonesia for US$275 million, and 5.56 per cent interest in Northwest Shelf Venture's oil fields in Australia for US$320 million. The other two oil companies also made large acquisitions, including CNPC's 30 per cent interest in two oilfields in Azerbaijian for US$52 million, SINOPEC's acquisition of an oilfield in Algeria for US$394 million, PetroChina's acquisition of six oilfields from US interest-controlled Devon Energy in Indonesia for US$216 million.

Several points need to be made clear: one is that the oil and gas deals made by the Chinese companies were only part of the broader 'going out' strategy. By the end of 2002, Chinese firms had set up almost 7,000 businesses abroad, with combined contractual investment exceeding US$16 billion. The business volume of Chinese firms through contracting engineering projects and overseas labour co-operation totalled US$97.2 billion (Lan, 2003: 24). Mergers and acquisitions (M&A) particularly had become the main form of China's direct investment abroad. The value of cross-border M&A purchased by Chinese companies increased from US$60 million in 1990 to $1.04 billion in 2002, a 17-fold increase. In sharp contrast to the remarkable decrease of M&A purchases by Japan, South Korea, and Taiwan since the mid-1990s, by 2002, China had become the fourth largest player in Asia next to Japan, Hong Kong and Singapore. To the Chinese government, this was a necessary and inevitable step toward economic globalization with its firms 'participating in international economic and technological cooperation and competition on a larger scale, in a wider field and at a higher level' (Lan, 2003: 24).[7] Moreover, multinationals are becoming powers in controlling the international petroleum market. They are expanding in the world's major exporting/importing countries via mergers and acquisitions and the development of upstream and downstream supply channels. These multinational giants have allied with international financial consortiums to emerge as major players, influencing international markets and international politics (Development Research Centre of the State Council, 2003). Consequently, if China wants to ensure some degree of oil security, the Chinese oil companies must join the ranks of these multinational giants.

Second, as some China watchers have acknowledged, oil and gas deals created a more successful image of Chinese companies than is actually the case. For example, 'in 2003, Chinese state-owned oil companies pumped 0.22 million b/d [barrels a day] of equity oil' (Logan, 2005) while its total oil import was over 5.6 million b/d. In early 2003, the overseas oil reserves controlled by China through its equity purchases accounted for only about 5 per cent of its feasible reserves at home (Zhai, 2003: 19–20). Even IEA predicts, with the current speed of investment from Chinese oil companies, China would be able to control only 1–2 per cent of global output by 2020. Clearly, Chinese oil companies did not invest in overseas equity for the immediate return of oil supply. Rather the price they paid can be seen as a sign that they were getting into the producing countries because they wanted to get a seat at the table for future international oil and gas distribution.

Third, much of the Chinese oil and gas acquisition was in the places where traditional importers were not present, such as Sudan, Angola, Venezuela, Thailand, PNG, etc. This is often seen as China being provocative – that is, China has taken advantages of the power vacuum in some of these countries and established its presence so that it could undermine US policies. These developments, however, can also be interpreted differently – China went to the countries where the US either had withdrawn, such as Sudan, or had had a very weak presence out of a desire not to provoke the US. Some Chinese have realized that 'to build an all-around well-off society internally and to maintain world peace and promote common development externally', China needs 'a peaceful Sino–US strategic relationship' (Wang, 2004b: 27). Moreover, China has made it clear that even though the Middle East is an area that enjoys the world's largest oil exports, exerts greatest influence and possesses enormous wealth, it is not the ideal place for China to get its oil because of its prolonged instability.

Fourth, investing in overseas oil production facilities is a two-way street. While China is interested in expanding its presence in oil-producing countries, the countries where China's oil companies invested have also welcomed investment for various reasons. When the two governments signed agreements that would allow Chinese companies to explore for oil and set up refineries in Venezuela, which is the fourth largest oil supplier to the US, Venezuela's president Hugo Chavez said that his country was seeking to reduce its dependence and would therefore like to give China greater access to Venezuelan natural resources. 'We have been producing and exporting oil for more than 100 years but they have been years of dependence on the United Sates', Chavez said. 'Now we are free and we make our resources available to the great country of China' (IAGS, 2005).

A similar view was held by the Canadian oil companies when the Chinese decided to invest in building a pipeline from Alberta to Vancouver so that it could import Canadian oil: the US has taken Canadian oil and gas for granted, 'the China outlet would change our dynamic. Our main link would still be with the US but this would give us multiple markets and competition for a prized resource' (Romeo, 2004: A1). Middle East oil-producing countries have held similar views that OPEC countries need China as a market and will make sure of its reliable supply. By securing a China that has 'limited political ambitions' as a customer such states can balance US influence and calm domestic opposition (Ogutcu, 2003; Calabrese, 2004; Bahgat, 2005). In Central Asia, countries that have oil reserves want to develop them to revive their economies, but they need foreign companies to help tap their rich reserves. Chinese efforts through the Shanghai Co-operation Organisation to settle the boundary disputes and to create a mechanism to promote confidence and security in the region paved the way for the Chinese inroad to the oil and gas development in the region (see Chung, 2004; and Goldstein, 2001).

National strategic oil reserves

Building a national strategic oil reserve is a relatively new idea for the Chinese. The idea of building China's strategic oil reserve merged in 1997 when CNPC and Sinopec, just before the reorganization, learned an expensive lesson – insufficient commercial storage capacity prevented them from taking advantage of low world oil prices and also forced them to cut production. In 1998, SDPC identified some possible storage sites and a year later four were suggested to the State Council (Downs, 2000; IEA, 2000). The proposal to build a national strategic oil reserve was confirmed in China's tenth five-year plan (2001–2005) to construct and use strategic oil reserves by 2005. China then was the only major oil-importing country without a strategic oil-reserve system. After setting it as an objective, the Chinese government sought help from the IEA to build strategic reserves from 2001 onward and it also allocated a little less than US$2 billion for the project. Different views have been expressed on whether China should build strategic oil reserves and if so, how. Those who supported the agenda argued that without reserves, Chinese industries would be vulnerable to the fluctuation of world oil supplies and its price changes. The government could intervene occasionally, but 'it is impossible for government to interfere for a certain industry or enterprise every time' (Tang, 2003: 38). China needed to build reserves to secure the safety of national energy resources. Others disagreed and argued that it would be too expensive to build oil reserves that would never meet even temporary demand during a crisis. For now, they argue there are so many other areas that demand government funding; building oil reserves that might not work is a waste of money. A third group agrees with the importance of building and holding oil reserves, but question the feasibility of doing so. Their main concerns are 'the amount of oil needed, when to start building the reserve and, more importantly, how to pay for it' (Zhai, 2003: 19). Having realized that whatever reserves China may be able to accumulate will not last long, some China watchers speculate that 'China may be more inclined to use strategic stocks to influence prices even without the threat of severe supply disruptions' (Logan, 2005). In sum, the debate continues – over the stockpiling strategy, the amount of oil needed, the time to start building reserves and, more importantly, who is going to pay.

Energy and security

Is China a threat to regional international stability as the result of its increasing energy demand? In addition to direct competition with the United States, there are several issues that can trigger conflicts – safety of sea-lanes for transporting oil, conflicting territorial claims in the South China Sea and East China Sea, China's new relationship with the countries in the Middle East, China's support for rogue states, and China's relationship with Central Asian oil-producing countries. It is argued in this paper that it is unlikely that China will engage in direct conflicts with major powers over these issues just for energy alone.

The Straits of Malacca

Like all its Asian neighbours, China worries about the security of the transport corridors, chiefly the Straits of Malacca, through which almost half a million ships pass each year. Some suspect that concern for safe transport motivated the Chinese government to ratify the UN Law of Sea Convention that guarantees safe passage. Others argue that since all of China's imported oil is delivered by tankers, safeguarding these deliveries means protecting sea-lanes and that is why China has been building up its naval capacity to cover the 7,000 miles of sea-lanes that lie between Shanghai and the Strait of Hormuz. Two points draw our attention: one is that China is not nearly as vulnerable as other oil-importing countries in Asia, particularly Japan and South Korea where over 90 per cent of oil consumption is satisfied with oil imports. Second, building up its naval capacity in China has been motivated more by political and diplomatic reasons other than concerns relating to oil imports, which after all accounts for less than 6 per cent of total oil consumption in China. This can be confirmed by the military experts who have been watching China's military developments closely: China's transformation of its military focus from a 'continental' defence to a 'peripheral' defence of the coastal and maritime regions 'is a result of a changing and uncertain strategic security environment within Asia' rather than from an internal drive to secure oil imports or for regional domination (Pultz, 2003).

South China Sea

There are plenty of discussions on the disputed territorial claims over the control of the Spratly Islands in the South China Sea. To some, 'the logic of energy security is based on an interrelationship of resources scarcity and territorial dispute that leads to conflicts' (Manning, 2000b: 86). The region is particularly problematic because of the nexus of territorial disputes and potential energy stocks. Three main issues are involved in the conflicts over the Spratly Islands. First, most analysts focusing on security issues tend to overestimate the oil reserves in the South China Sea. 'No one really knows how much oil is under the Spratlys [and] geological estimates vary widely' (Fesharaki, 1999: 92). What we do know is that between 1980 and 2000, with $5 billion investment, the oil extracted from the area accounted for less than 1 per cent of China's domestic consumption. Second, given the limited knowledge of oil reserves and limited supplies so far, it is clear that the conflicts are more about sovereignty – which remains a crucial concept in international politics – military presence and political leverage. 'Oil is being used as an excuse or justification to try to assert power' (Fesharaki, 1999: 92). Third, because it is a political issue rather than an energy one, all sides at different points show their willingness to compromise. The examples are the *Declaration on the Conduct of Parties in the South China* Sea signed in 2002 to reduce tension and the *Code of Conduct in the South China Sea* signed between China and ASEAN in November 2002, committing all signatories to peaceful resolutions of standing disputes. These agreements may not have the binding force many people in the West prefer to see. This, however, should neither be a surprise, nor be taken lightly. Informal forms of diplomacy in Asia are more a norm than an exception (see Pdgaard, 2002; and Tonnesson, 2000).

China and Japan

Two issues concerning energy are the territorial dispute over the Diaoyu Tai (Senkaku) islands and the gas pipeline from Russia. Similar analyses can be made about the disputes between China (with Taiwan) and Japan in the East China Sea, under which is supposed to be gas-rich areas. At issue is each country's claim to its exclusive economic zone under the Third UN Convention on the Law of Sea. The two countries have never agreed on a maritime border and the Daiyu Tai islands are in the middle of the disputed areas – 168 kilometres north-east of Taiwan and 406 kilometres west of Okinawa. When China started test drilling in the Chunxiao gas field, lying in waters that both sides agree belong to China, Japan protested, worrying that China's drilling would siphon off gas buried under the seabed on the Japanese side. When the Japanese government granted exploration rights to the Teikoku Oil to exploit the area's fuel reserves for test drilling, the Chinese government protested. While the western media predicted conflict between China and Japan, the government of both sides engaged in endless and seemingly fruitless negotiations.

The second territorial dispute concerns the construction of pipelines from Siberia. In July 2001, Chinese leader Jiang Zemin visited Russia and signed several important energy-trade agreements. One of these called for the feasibility study of a pipeline from east Siberia to eastern China. China preferred to have a pipeline built between the Russian city of Angarsk and Daqing, the major oil field in the north-east of China while Japan pushed for a route from Angarsk to Nakodka (Russia's Pacific coast port). In 2002, Russian officials pledged $2 billion to build a 2,247-km pipeline from Angarsk to Daqing, which was scheduled to begin in 2003 and commissioned by 2005. The construction and operation of this pipeline would involve Russian Transneft and Yukos and Chinese Sinopec and CNPC. Bitter competition ensued over the next 2 years involving not only two countries – China and Japan – but also two Russian companies – the out-of-favour Yukos that preferred the shorter route to China and Transneft, the state oil pipeline company that supported the much longer and more expensive route to the port of Nakodka, arguing that it would allow Russia to serve several markets and maintain control over its export options.

In March 2004, Russia proposed the third route, Taishet–Nakodka that would be 4,130 km long – some 250 km longer than the Angarsk–Nakodka – and projected to cost between $11 to $16.5 billion, nearly three times the cost of Angarsk–Nakodka, which the Japanese government originally proposed. It was reported that Japan had promised up to $14 billion funding for the pipeline and additional $8 billion in investments in the Sakhalin-1 and Sakhalin-2 oil and gas projects. The Japanese investment would come from its state-owned Japan National Oil Corp. When the Russian Prime Minister announced the decision to go ahead with the Taishet–Nakodka project late 2004, the western media reported that 'Russia snubs China over pipeline route', 'a real war broke out in the Northeast Asia', or the 'struggle for Angarsk' was on 'a plane with the Iraq War' (Christoffersen, 2004: 3).

Meanwhile, Chinese officials seem to 'have remained diplomatic' and realized that Russian domestic politics was the main factor in the final decisions (Blagov, 2004; Christoffersen, 2004; Bahgat, 2005). Instead of confronting the Russian or Japanese government, China seems to have accepted the explanation offered by the Russians that a branch of pipeline would be built on the route to Japan and started negotiating the possibility of importing electricity directly from the Irkusk. What this shows is that it is important for China to get access to energy supplies, not only oil and gas but in all forms. Yet, it does not want to do so at the expense of its relationships with any of the major powers or regional or world stability. After all, as Qian Qichen, a Chinese veteran diplomat, once commented on 'the China threat', 'if China's comprehensive power today were at the same level at it was a decade ago, there would be no such loud voices about the "China threat", and there will not be market for this theory in a few decades from now on when China becomes much more developed' (Wang, 2004a).

Central Asia

The Central Asia region has vast oil and gas resources, particularly in Kazakhstan, Uzbekistan and Turkmenistan. In June 1997, CNPC purchased 60 per cent of Kazakhstan's Aktyubinsk Oil Company for $4.3 billion. It also announced the construction of a 3,000-km pipeline linking western Kazakhstan and China's Xinjiang region with a price tag of $3.5 billion. In total, 'between early 1997 to early 2005, CNPC invested nearly $10 billion in Kazakhstan' (Xu, 1999; Wu & Han, 2005: 18). Many have taken China's interests and investments in the region as a potential threat to the interests of the US and consequently to regional stability. The argument goes: China took advantage of the absence of the US military in the region and the deals allowed China to avoid the sea-lanes dominated by the US Navy and to pass through the regions where China's land power has the advantage. In doing so, the US dominance in the region is severely constrained – one wins; the other must lose (Downs, 2000).

Several issues need to be clarified. First, even though there are plenty of oil and gas reserves in Central Asia, transportation will be a major problem since they are after all land-locked countries. It would make much more economic sense if oil is transported to the Mediterranean and then shipped through the normal sea-lanes (Ogutcu, 2003). This fact disputes the argument provided above. Second, the pipeline from Central Asia to Tarim in Xinjiang would not do any good unless and until the Chinese government launched the program of 'developing the West' of which constructing a pipeline between Tarim and Shanghai is an important component. CNPC raised the issue of constructing a west-east pipeline in 1996 but it was not endorsed until 2000 when the central government decided to pull the country out of the economic slow-down with an increase in public spending (Xu, 1999). In other words, it was not considerations of securing energy access in Central Asia that made the project possible. It was domestic politics. Energy policy toward Central Asia was in essence a surrogate for foreign policy issues – preventing Muslim radicals from spreading to the western part of China and maintaining stability in the region (Chung, 2004).

Conclusion

China's energy demand is undoubtedly on the rise and it is a challenge to Chinese policy-makers as well as the international community as a whole. Three issues need to be highlighted: first, energy security is more a domestic political problem than a foreign policy issue in China. Creating an effective integrated central authority is of central importance to translating the energy strategies to policies. Switching the emphasis from increasing energy supply to demand management, especially energy conservation, is the right policy, yet it is difficult to implement not only because of existing interests in the energy sector but also because of the broader economic and social impacts as the result of the switch. Second, foreign energy policy is foreign policy first and energy second. This interpretation of the Chinese energy policy allows us to place the energy policy in a broader political and diplomatic context and then emphasize China's desire to be recognized as a major global player. Even Chinese policy-makers have realized that there is a long way to go for China to become a major player and therefore maintaining a peaceful domestic and international environment is essential for its development. It is unlikely that China would get into direct confrontation with any major power over the control of energy resources. Finally, analysing China's energy security by placing its policy in a broader context does not mean that there is guaranteed peace and stability. It does, however, raise the possibility of co-operation, as many Europeans have realized:

The public discourse in the United States concerning China invariably refers to its rise and is dominated by analysis of China's increasing hard power: the growth in China's military power and its effect on US national security interests in East Asia ... This is the principal prism through which most US analysts view China's rise and the main factor that animates the debate in Washington. ... Europe, on the other hand, considers China's rise more in terms of China's domestic transitions, that is, Europeans see China as a large developing country in the midst of multiple transitions leading it away from state socialism and toward a market economy, a more open society, and a more presentative and accountable government. ... This perspective underlies the main thrust of European policies toward China: to assist China in successfully managing these internal transitions and reforms. Europe does not want China to become a failed state [and] is willing to accept China as it is and to assist Beijing in meeting its domestic challenge.

(Shambaugh, 2005: 14–15)

Notes

1 For the European view on the rising China, see Shambaugh (2005: 7–25). For the China's demand as a threat to the international stability, see Myers and Lewis (2002: 115–134); Bajpaee (2005b) and Haider (2005).

2 Based on official figures, China's primary energy demand grew at over 5 per cent a year between 1981 and 1995 while its annual GDP growth rate was about 9.8 per cent (China Statistical Bureau, *China Statistical Yearbook*, various years). Other studies have downgraded the GDP growth rate to about 7.6 per cent and this has significant implication to energy intensity (see Sinton & Fridley, 2000: 671–687; and IEA, 2000).

3 National pride has always been an important factor in the development of petroleum industry in China. It is also part of the recent drive for Chinese oil companies to establish themselves as world-class corporations.

4 After the Asian financial crisis in 1997–1998, there was an unexpected decline in energy demand. The government made a decision not to build new conventional coal-fuelled plants for 3 years and at the same time, it pumped huge amounts of capital to boost the economy to deal with aftermath effects of the Asian financial crisis. As the economy recovered from the slow growth, energy-intensive industries in particular expanded. The combination of these policies led to a national electricity supply shortage.

5 Particularly studies published by the James A. Baker III Institute for Public Policy of Rice University and Rand Corporation.

6 See CNOOC website on international co-operation. http://www.cnooc.com.cn.; and Li (2003).

7 See also UNCTAD, *World Investment Report*, various years.

5 Contours of India's energy security: harmonising domestic and external options

Ashutosh Misra

Pressing concerns about energy security have pushed states across the world into a resource search frenzy, and India is no exception. All necessary means are being adopted ranging from amending existing domestic polices wherever required, to exploring viable options overseas. India, being the world's sixth largest energy consumer in the wake of robust economic development, industrialization, rising population and rapidly growing numbers of vehicles on the road, needs a comprehensive energy security regime. Steadily a realization has set in that the state alone would not be able to effectively address the challenges energy security poses, and therefore needs to engage the private sector as well.

In this context India is continuously reviewing and amending its existing energy policies and also searching for alternative supplies overseas. The public and private sector companies through consortiums have had their tasks carved out in exploration, refining, marketing and transportation, with a fair degree of success. Still, a lot remains to be achieved.

India's dependence on oil supplies from the Gulf makes it vulnerable to oil shocks and there is a pressing need to diversify its energy imports. With rocketing oil prices India like other states is consciously looking to natural gas as an alternative to oil, which is both cost-effective and 'Kyoto' friendly. The perceived perils of energy shortages by 2025 are pushing states to stretch out in feasible directions and back home searching for available resources and renewable energy sources. India has been energy hunting in West Asia, the Persian Gulf, Central Asia, South Asia, and the Asia–Pacific with mixed results. The search for energy in the Asia–Pacific has gained momentum in the last 4–5 years. South-East Asia and the Asia–Pacific could potentially provide some relief to India's growing energy needs. This chapter deals with the contours of India's energy concerns encompassing the domestic situation, official policies and strategies and overseas ventures.

India's energy scenario

India is the sixth largest energy consumer in the world and the third largest producer of coal. Being a fossil fuel-deficient country, having scanty oil reserves and poor-quality coal, India's energy situation solicits serious consideration. The following statistics give an estimate of the gravity of challenge that India confronts in its energy sector.

Oil

As per estimates, oil consumption worldwide will rise from 78 million barrels per day in 2002 to 103 million by 2015 and 119 million by 2025. The figures in the International Energy Outlook (IEO) 2004 are a trifle higher than in IEO 2005 (121 million barrels) (EIA, 2004: 8), which indicates, perhaps successes on part of states in addressing the problem.

Around 30 per cent of India's energy requirements are met through oil. Currently around 4.5 billion barrels (bbls) of reserves are located in Mumbai High, Upper Assam, Cambay, Krishna–Godavari and Cauvery Basins. The largest Indian oil field is located in the offshore Mumbai High field and its current production stands at 260,000 barrels per day (bl/d). In 2003, the total oil production was 819,000 bl/d including 660,000 bl/d of crude oil production (EIA, 2005e). India has proven reserves to the tune of 5.4 bbls, reserve growth (reserve increase due to technological factors that enhance recovery) of 3.8 bbls, and undiscovered reserves around 6.8 bbls (EIA, 2005d). India along with China and other 'emerging economies' in Asia is projected to experience a combined economic growth of 5.5 per cent annually between 2002 and 2025, which would translate to 3.5 per cent annual increase in regional oil use (EIA, 2005d: 27; Tables 5.1 and 5.2).

Ministry of Petroleum and Natural Gas (MPNG) statistics reveal that India presently imports around 69 per cent of its oil, of which 67 per cent comes from the Middle East region. India's crude oil production during the year 2003–2004 was 33.38 million metric tonnes (MMT) and gas production was 31.95 billion

Table 5.1 Estimated energy demand (the X five-year plan (2002–2007), Planning Commission, India)

Primary fuel	Demand in original		Demand (MTOE)	
	2006/2007	2011/2012	2006/2007	2011/2012
Coal (MT)	460.50	620.00	190.00	254.93
Lignite (MT)	57.79	81.54	15.51	22.05
Oil (MT)	134.50	172.47	144.58	185.40
Natural gas (BCM)	47.45	64.00	42.70	57.60
Hydro power (BkWh)	148.08	215.66	12.73	18.54
Nuclear power (BkWh)	23.15	54.74	6.04	14.16
Wind power (BkWh)	4.00	11.62	0.35	1.00
Total commercial energy	–	–	411.91	553.68
Non-commercial energy	–	–	151.30	170.25
Total energy demand	–	–	563.21	723.93

MT = million tonnes; BCM = billion cubic metre; BkWh = billion kilowatt hour.

Source: http://www.planningcommission.nic.in/plans/planrel/fiveyr/10th/volume2/v2_ch7_3.pdf

Table 5.2 Primary energy production in India

Source	1970/1971	1980/1981	1990/1991	2001/2002	2002/2003
Coal and lignite (MT)	76.34	119.02	228.13	352.60	367.29
Crude oil (MT)	6.82	10.51	33.02	32.03	33.04
Natural gas (BCM)	1.45	2.36	18.00	29.71	31.40
Nuclear power (BkWh)	2.42	3.00	6.14	19.48	19.39
Hydro power (BkWh)	25.25	46.54	71.66	73.70	64.10
Wind power (BkWh)	–	–	0.03	1.97	2.1

MT = million tonnes; BCM = billion cubic metre; BkWh = billion kilowatt.

Compiled from various sources.

cubic metres (BCM). During 2003–2004, India imported 90.434 MMT of crude oil worth Rs. 83,525 crore. During the same period petroleum products worth Rs. 9,640 crore were imported and 14,620 MMT of petroleum valued at Rs. 16,781 were exported. India exported petroleum products to the USA, Singapore, Iran, Brazil, Malaysia, Mexico, Sri Lanka and Japan among others (Ministry of Petroleum and Natural Gas, 2004: 2–3). In 2004–2005, crude oil bills mounted to Rs. 117,032 crore for importing 95.9 MMT of crude oil (EIA, 2005e).

Natural gas

With the entry into force of the Kyoto Protocol on 16 February 2005, countries are gradually switching to natural gas to fuel their electricity and industrial sectors and reduce consumption of coal and oil. In the 2002–2025 period, the world's consumption of natural gas is estimated to increase by 69 per cent, from 92 trillion cubic feet to 156 trillion cubic feet, and its share of total energy consumption is projected to grow from 23 per cent to 25 per cent. Electricity generation accounts for 51 per cent and the industrial sector accounts for 36 per cent of the growth in world's natural gas demand between 2002 and 2025 (EIA, 2004: 3).

Since the mid-1990s, the consumption of natural gas in India has increased more than any other form of fuel. In 1995, the consumption of natural gas was around 0.6 trillion cubic feet (TCF) per annum which increased to 0.9 TCF in 2002 and is estimated to reach 1.2 TCF by 2010 and 1.6 TCF by 2015 (EIA, 2005e). Discoveries of nine gas reserves in the Krishna–Godavari (KG) basin, offshore from Andhra Pradesh in October 2002 was a significant breakthrough for Reliance Industries Limited (RIL). Current estimates suggest availability of 7 TCF feet of gas there. Another find by RIL came in June 2004 offshore from Orissa with an estimated reserve of 1 TCF. Cairns Energy discovered natural gas in 2002, offshore from Andhra Pradesh and Gujarat containing total gas reserves around 2 TCF. These findings would to a great extent be able to cater to the energy demands particularly in the east coast of India. However, India is still short of natural gas supplies and has to import much of it from abroad. Although

India's natural gas consumption will grow at an annual rate of 4.8 per cent, natural gas will continue to be a minor fuel constituting only 6.5 per cent of total primary energy consumption. India is heavily investing in infrastructure expansion to facilitate domestic transportation and import of gas (EIA, 2004: 43).

Coal

India is the third largest producer of coal after China and the United States and it is predominantly used in India due to its abundance and cheaper exploration and extraction costs. According to the IEO 2005, coal's share of total world energy consumption is around 24 per cent (2002 figures) and in the electricity sector it is expected to reduce from 39 per cent in 2002 to 38 per cent in 2025, unlike in the industrial sector where it will rise from 20 per cent in 2002 to 22 per cent in 2015, and stabilize through 2025 (EIA, 2004: 49).

Among the 'emerging economies' in Asia coal consumption is projected to double from 2,118 Mmst in 2002 to 3,715 Mmst in 2015 and 4,435 Mmst in 2025 and India and China will be major countries raising the coal consumption share from 40 per cent in 2002 to 51 per cent in 2015 and 54 per cent in 2025. Both account for 71 per cent of the total increase in coal consumption worldwide due to their higher economic growth. The IEO 2005 shows that the energy sector in India accounts for less than 60 per cent of coal consumption, with the industrial sector accounting for the remaining 40 per cent. In the electricity sector coal consumption is projected to rise by 2.2 per cent annually, from 5.1 quadrillion British Thermal Units (Btu) in 2002 to 8.5 quadrillion in 2025. By 2012 India has targeted construction of around 40,000 megawatts of new coal-fired generating capacity (EIA, 2004: 53–54). The problem in the use of coal is that the Indian variety is not of a high quality and contains high ash content and low calorific value, so most cooking coal has to be imported. Most of the 390 coal mines under the state-controlled Coal India Limited (CIL) are located in Bihar, West Bengal and Madhya Pradesh which provide 90 per cent of the coal used for domestic consumption (EIA, 2005e).

Electricity

Robust economic development, booming population, rapid urbanization and industrialization demands more electricity. In India and China, household energy consumption is estimated to increase in the next 10–20 years and the residential sector demand bulk of the electricity supply. India is currently seventh greatest electricity consumer in the world and is ranked sixth in terms of the total installed electricity generating capacity which translates to 3.3 per cent of the world's total (EIA, 2005e).

Although the electrification drive has gained momentum in India, at the dawn of the twenty-first century, around 45 per cent of households still do not have access to electricity. Plus, there are problems of power failures, aging equipment, power theft, blackouts and brownouts and tardy redressal mechanisms for consumers, which are creating problems for both consumers and suppliers. The state electricity boards (SEBs) are facing a financial crunch due to lack of funds, poor infrastructure and corruption. High subsidies on electricity in states like Punjab add more to the problem and the SEBs cannot even recover their basic costs. Efforts are being made to reduce cross-sector subsidy burdens to minimize the losses of SEBs. In 2002, India's SEBs suffered a loss of $5.3 billion.

From 1998 the Indian government has encouraged foreign investment and allowed foreign equity from 74 per cent to 100 per cent in power generation and distribution from hydroelectric, coal, lignite, oil, or gas power plants – the exception being with nuclear plants (EIA, 2005e). The Electricity Act 2003 was passed to address the problems of the SEBs but still large foreign investments in the electricity sector are yet to come. In March 2005, India announced plans for subsidized electricity consumption for rural and poor households, which may begin to show effects in the coming years (EIA, 2004: 67).

Downstream/refining performance

India has been importing refined products in large quantities until 1999, when construction of refineries started. By 2003 India had a total of 2.1 million bbl/d in refining capacity, an increase of 970,000 bl/d since 1998. Reliance Petroleum's refinery based in Jamnagar, Gujarat, which currently has reached its peak capacity at 540,000 bl/d, was the largest addition in the refinery sector in 1999. Its products are being marketed by the state-owned entities but for the future Reliance plans to achieve self-sufficiency in transporting and marketing (EIA, 2004: 67).

India's domestic refining capacity on 1 April 2004 stood at 125.97 million metric tonnes per annum (MMTPA). The MPNG Annual Report shows that the output of petroleum products during 2003–2004 from the domestic refineries and non-refineries was sufficient to meet all domestic demand except for Liquefied Petroleum Gas (LPG) (Ministry of Petroleum and Natural Gas, 2004). There are several other companies in the refinery sector including Oil and Natural Gas Commission (ONGC) and Indian Oil Corporation (IOC) (Ministry of Petroleum and Natural Gas, 2004: 26–30).

India is expanding its pipeline network for the transportation of petroleum products. In developed countries around 60 per cent of total petroleum products are transported through pipelines; in India the proportion is around 32 per cent and is estimated to go up to 45 per cent in the next 4–5 years (Ministry of Petroleum and Natural Gas, 2004: 26–30).

In November, 2002 the Indian government called for expressions of interest for laying down pipelines, which resulted in companies locking horns with each other and questioning the Government's draft policy of September 2003 (Ministry of Petroleum and Natural Gas, 2004: 26–30). Gas Authority of India Limited (GAIL) opposed Reliance's bid to build the pipeline from the KG basin in Andhra Pradesh to Gujarat arguing that GAIL is the notified company for building pipelines ('Gail Bid to Thwart Reliance Pipeline', 2002). The Ministry has to iron out these differences and lay down clear guidelines for pipeline construction.

India's energy security strategy

India has been making serious efforts in formulating a comprehensive and effective energy policy over the past decade or so. MPNG has effected changes in energy polices through consultation with private and public sector entities and brought about reforms to instil vigour in energy security efforts. The awareness of India's energy challenges over the years has increased and been integrated into India's larger energy strategy. A sense of urgency can be seen in the government's plans, and support for putting a forceful energy strategy in place has come from the highest offices.

On 14 August 2005, President of India, A.P.J. Abdul Kalam, in an address to the nation said, '... Today on this 59th Independence Day, I would like to discuss with all of you another important area that is "Energy Security" as a transition to total "Energy Independence"' (Kalam, 2005). He laid down two key principles of energy security. First, to use the least amount of energy to provide services and cut down energy losses. And second, to secure access to all sources of energy including coal, oil and gas supplies worldwide, until the end of the fossil fuel era, thought by some to be fast approaching. Simultaneously, India should access technologies to provide diverse supplies of reliable, affordable and environmentally sustainable energy (Kalam, 2005).

Considering India's current energy predicaments and long-term security challenges, an integrated strategy which synergizes its domestic and external options has become the cardinal requirement. Such a long-term energy strategy should consist of first, pursuit of access to sources of oil and gas abroad; second, locating domestic reserves and enhancing capacity to explore, produce and transport; third, promoting public awareness and participation in energy conservation; and fourth, developing of the renewable energy (RE) sector.

India hydrocarbon vision-2025

In March 2000, former Prime Minister Atal Behari Vajpayee placed the India Hydrocarbon Vision (IHV)-2025 Report in both Houses of Parliament (Ministry of Finance, 2000: 139). The IHV lays down the medium and long-term policy of the government in the energy sector covering the entire gamut of issues including exploration, refining, marketing infrastructure, gas and all other related matters in the hydrocarbon sector. The report has been prepared by a ministerial group appointed by the prime minister to 'assure energy security by achieving self reliance', increase 'indigenous production' and attract 'investments'; enhance 'quality of life' and ensure 'cleaner and greener' India; and develop the hydrocarbon sector as a 'globally competitive industry'.

The X Plan (2002–2007)

Consequently India's Tenth Plan brought about a paradigm shift in existing polices and called for international competitive bidding in a deregulated context;

targeting the appraisal of 35 per cent of India's total sedimentary basins; acquiring acreages abroad and introducing advanced technologies for exploration, refining, and enhancing transportation. The plan envisages optimizing production of natural gas and oil from domestic basins and existing fields; synthesizing intensive exploration and production and increasing the reserve base in frontier areas and deep waters; improving recovery rates by 3–4 per cent in major fields; stressing quality of exploration; undertaking projects and studies in producing fields for enhanced recovery; acquiring equity oil abroad; and exploring of Coal Bed Methane as an alternative fuel (Ministry of Petroleum and Natural Gas, 2004: 12). The MPNG Minister Mani Shankar Aiyar inaugurated as the first of its kind, Centre for Research on Energy Security at The Energy and Resources Institute (TERI) with the participation of diverse stakeholders – different government bodies, corporates, non-governmental organizations (NGOs) and experts from various fields. The centre will focus on international developments, geopolitical trends, and international co-operation with regards to energy security and bring out quarterly reports (Mani Shankar Aiyar inaugurates TERI Centre for Research on Energy Security, 2005).

ONGC–Videsh Limited (OVL) has taken the lead in the pursuit of acquiring exploration acreage and oil/gas-producing properties overseas. It has already acquired or discovered a 45 per cent share in gas-producing properties in Vietnam, a 20 per cent share in oil and gas properties in Russia and in a 25 per cent stake in Sudanese oil fields. From Vietnam and Sudan India's production share is around 7.54 million metric standard cubic meters per day of gas and 2,50,000 barrels of oil per day, respectively. The shipment of oil already commenced from Sudan in May 2003. In Russia, the Sakhalin Project is under development and OVL is currently exploring possibilities of blocks in Iran, Myanmar, Iraq, Libya and Syria. In Myanmar an exploratory well was drilled and natural gas was discovered in January 2004 (Ministry of Petroleum and Natural Gas, 2004: 2).

New Exploration Licensing Policy (NELP)

NELP was first mooted in 1997 to permit private/foreign involvement in technology exploration and production. It provides an international class fiscal and contract framework for exploration and production of hydrocarbons. After the operationalization of NELP over the first three rounds, production sharing contracts (PSCs) for 70 blocks had been signed and under round four 24 blocks were offered, out of which contracts on 21 blocks were signed on 6 February 2004. According to the MPNG annual report, in the first four rounds of NELP from 2000 to 2004, contracts for 91 blocks covering about 9.0 lakh sq. km had been assigned and the area under exploration went up to 12,40,000 sq. km. Currently around 74 per cent of the area under the exploration and production belongs to the NELP blocks. The total investment in these 90 blocks in three phases, would be to the tune of Rs. 19,050 crore and until September 2003 over Rs. 3,000 had already been invested by private/joint and national oil companies. The dividends from the discoveries of hydrocarbon reserves have already begun to accrue. The discovery of the KG basin in October 2002 is the case in point (Ministry of Petroleum and Natural Gas, 2004: 2).

Strategic storage

For a country like India whose dependence on imports is around 69 per cent out of which 67 per cent comes from the Middle East, maintaining a strategic reserve for emergencies is indispensable. In January 2004 the Indian government announced the development of 5 MMT of strategic crude oil storage at various locations. This storage will add to the existing storage of crude oil and petroleum products by oil companies and will provide an emergency response mechanism during crises. The storage facility would be managed by a special purpose vehicle, fully owned by one of the oil PSUs.

Upon completion of this facility in around 2008, India will qualify for membership of the International Energy Agency wherein the members are committed to hold a stock equivalent to 90 days of net oil imports. The total cost of the storage is calculated around Rs. 1.650 crore, requiring annual operation cost of Rs. 40 crore and for storing 5 MMT of crude oil worth Rs. 5,000 crore.

National auto fuel policy

As part of its firm resolve to reduce oil consumption in the transportation sector and also reduce emission levels the Indian government formulated a comprehensive national Auto Policy on 3 October 2003. This policy encapsulates regulating vehicular emission norms such as Bharat Stage II standards in the country with effect from 1 April 2005 and Euro III equivalent with effect from 1 April 2010 for passenger cars, light commercial vehicles and heavy-duty diesel vehicles. Bharat II standards would come into effect from 1 April 2008 or not later than 1 April 2010 in metropolises and major cities (Ministry of Petroleum and Natural Gas, 2004: 2).

While the use of liquid fuels continues, use of Compressed Natural Gas (CNG) and LPG would be encouraged. Besides, the development of alternative fuel vehicles such as battery, hydrogen and fuel cell operated would be accelerated. In cities with higher vehicular pollution, vehicle owners have to get a pollution under control certificate at regular intervals as well. Delhi has become the first city where all commercial vehicles have been converted to run on CNG which has significantly reduced emission levels. The MPNG in synergy with the concerned ministries has formulated necessary legislative and regulatory framework for safe usage of LPG as an automotive fuel.

Chasing pipeline dreams

The resolve to gradually switch over to gas from liquid fuel, has pushed India to bring in more gas through imports to meet the fuel demands in core sectors like power, fertilizer and steel. The government has established Petronet LNG Ltd. (PLL), a joint venture of IOC, ONGC, GAIL, and Bharat Petroleum Corporation Limited to chase all viable possibilities of LNG supplies from abroad, especially in Iran, Myanmar and Turkmenistan. PLL already commissioned 5 MMTPA terminals at Dahej, Gujarat in February 2004 and pipeline construction has gained pace since March 2004 (Ministry of Petroleum and Natural Gas, 2004: 8).

An in-principle agreement was reached in May 2003 with Iran to boost cooperation in the hydrocarbon sector to enable import of 5 MMTPA of LNG in lieu of Indian Oil PSUs' participation in a discovered and semi-discovered oil and gas field. The PSUs would also help upgrade Iranian refineries and petrochemical plants.

Iran–Pakistan–India pipeline

The 2,670-km long pipeline from Iran is to pass through Pakistan and given the unpredictable nature of Indo–Pakistan relations, security of the pipeline has always been India's prime concern. The other problems confronting the pipeline have been lack of financers, and the reported US reservations over the $4 billion gas pipeline due the US–Iran nuclear impasse.

Notwithstanding the above impediments the three countries nonetheless have pushed the project with immense determination and conviction. Recently, as per press reports, Italy's ENI company, the sixth largest in oil and gas production in the world has expressed its willingness to finance the project and some Indian firms have also followed suit. This is a significant development. The pipeline would run for about 1,115 km through Iran, 705 km through Pakistan and 850 km through India (ul-Haque, 2005). Pakistan would be required to invest $1 billion and would be entitled to $150 million as transit fees annually from India for allowing the supplies to run through its territory. Latest estimates show that due to rising steel prices the cost of the pipeline may go up by 75 per cent pushing the cost up from $4 billion to $7 billion ('Pakistan, India play pipeline tune', 2005).

Myanmar–India pipeline

A Myanmar–India pipeline is another option available for gas transportation to India. The possibility gained vigour after a massive gas reserve was found in A-1 block in the Rakhie basin in Myanmar by an Indian–Korean consortium to the tune of 4.2 TCF to 5.8 TCF. GAIL has already been offered a 65 per cent share by Myanmar in production, in addition to two new blocks in A-2 and A-3 on 'nomination' basis. GAIL is presently exploring overland and undersea pipeline routes from Myanmar. The third option is to bring the pipeline through the northeast of India, which is a dangerous proposition due to an ongoing insurgency (Mangla, 2004: 16). The most viable option, it seems, would be to negotiate with Bangladesh, with which India's relations have not been very smooth in recent times.

Bangladesh–India pipeline

India has also drawn plans for a possible pipeline from the Bibiayana fields in north-east Bangladesh to Delhi. The project is a 'low probability' venture and is still pending approval in the Bangladeshi Parliament. IOC, GAIL and ONGC have already pitched in to buy the prospective gas from Bangladesh through a consortium called India International Gas Company. The main concern before Bangladesh remains whether it has enough reserves to meet its domestic demands and export also. There are political reasons as well, since India and Bangladesh do not enjoy a smooth relationship and any Bangladeshi party supporting the project would be labelled as pro-India and risk political reversals. Besides, the high cost of a pipeline for bringing from eastern Bangladesh to gas markets in central south and north India is also a factor (Mangla, 2004: 16). India nevertheless is keen to push this project with Bangladesh.

Turkmenistan–Afghanistan–Pakistan–India pipeline

India is also hopeful about the proposed 1,700-km long pipeline from Turkmenistan, popularly known as the Turkmenistan–Afghanistan–Pakistan pipeline to transport 20 BCM of gas per annum. An Asian Development Bank study estimated the cost to be around $2.5–$3.0 billion. The pipeline would carry gas from Dauletabad fields in south-east Turkmenistan to Afghanistan, Pakistan and possibly to India. A UNOCAL-led consortium had expressed its willingness to undertake the laying of 1,270 km of the pipeline at the cost of $2 billion from Dauletabad to Pakistan via Afghanistan and with a possible extension of 640 km from Multan to Delhi (Mangla, 2004: 16). For the project to be feasible, stability in the Afghan corridor is the key ('The Afghan Corridor', 2002).

New policy impetus

Prime Minister Manmohan Singh, an economist himself, understands well that for India's economic growth and development, energy security is the key. He has furthered all necessary support for the cause and in this connection agreed to preside over the first meeting of the high-power Energy Coordination Committee which looks at India's energy concerns. The government finds two pressing issues at hand: first, a 38 per cent shortage of natural gas for power generation and second, a sudden coal crunch with supply dipping by 7 per cent below requirements. On the coal front, the government has decided to enhance domestic supplies, to import coal at possibly higher rates and to acquire coal mines in Australia, Indonesia, Mozambique and Zimbabwe. By December 2006, the government targets aim at producing an additional 10 MT of coal from 10 coal mines and clear mining in another 11 mines which could fetch up to 81.5 MT of coal. The government has warned the coal companies that under-performance would lead to cancellations of their licenses. To minimize transportation cost plans are being made to build power plants in coastal areas where the coal can go straight from the ports ('PM draws up energy blueprint', 2005).

To boost natural gas supplies, the government will 'monitor' the development plans of Reliance and other private companies and encourage them to achieve full-scale production by 2008–2009. The MPNG is also pushing Qatar's RasGas to increase its supplies and is expediting an agreement with Russia for gas imports from the Sakhalin fields. In addition the Finance Minister P. Chidambram has hinted that the government may scrap the custom duty on LNG imports and ask state governments to lift sales taxes on liquefied gas ('PM draws up energy blueprint', 2005).

The government is also approving the 2,100 MW Dabhol power project in its new incarnation in which a consortium of Indian companies takes control from foreign promoters and lenders. The Cabinet is also likely to ease rules and conditions for approving new power stations. For projects costing up to Rs. 15,000 there will be no need for a clearance from the Union Cabinet Committee on Economic Affairs and this could be done directly with the Power Ministry's approval. For coal-based projects the deadline for approval is likely to be fixed at 24 weeks. For smaller hydel projects up to Rs. 500 crore, approval may be given at the power secretary level, whereas the present policy requires the cabinet's approval ('Manmohan Cabinet Goes on Energy Overdrive Today', 2005).

Analysts observe that India's energy policy and strategy rests on five pillars. First, there is a need to switch from oil to gas. Second, in the hydrocarbon sector, the NELP should be implemented vigorously. Third, there is a need for accurate price signals and encouraging competition to promote energy. Fourth, through a healthy mix of private and public effort, the domestic search for possible reserves should be pursued in a robust manner. And last, it is pertinent to diversify sources of energy to ensure a stable energy future immune from oil shocks and geopolitical manipulations (Malhotra, 2005: 56–57).

There are others who believe that the biggest challenge rests in bringing about major institutional changes. It involves establishing suitably empowered and totally independent regulatory commissions for pricing insulated from all pressures and interests. The Indian power sector is marred with technological and financial handicaps, meaning that private participation will consequently boost the distribution of power as well (Pachauri n.d.). Indian Prime Minister Manmohan Singh also emphasized the need to have an economic pricing policy cognizant of the social inequities and income disparities. This is vital for power conservation as well (Singh, 2004: 11).

India's energy future: options and challenges

Under the Kyoto Protocol India does not figure in the Annex I list of countries required to bring down or limit emissions of carbon dioxide and other green house gases over the first commitment period i.e. January 2008–December 2012, to a level that was determined as part of the negotiation process. India has ratified the Protocol and maintains the build-up of green house emissions has been the creation of developed countries and the developing countries cannot compromise on their development needs for want of the Protocol provisions. Driven by growing energy constraints, New Delhi has been exploring alternative sources of energy to supplement up to some extent the conventional fuels and the existence of a separate Ministry of Non-Conventional Energy Sources (MNES) underlines the importance of this sector.

RE

In the next 25 years, due to an increase in demand for fossil fuels, high costs, added pressure on Gulf-based fuel supplies and a rise in green house emissions, the share of RE sources in power generation ought to be augmented. India's experience and efforts in the RE field have not been very encouraging and the share of renewables is as low as 3 per cent or a mere 4,000 MW. Given the abundance of solar, wind, water and ocean resources there remains a vast untapped and unexplored potential in RE for India, which could be raised up to 10 per cent of the energy production by 2012. In India solar photovoltaic (SPV) technology has been rather successful largely funded by the Indian Renewable Energy Development Agency Ltd (IREDA). Home lighting systems, solar lanterns and pumps could be installed in large numbers in several states. Other renewable sources are biomass and bagasse-based cogeneration with a total potential of 20,000 MW ('Action not Word', 2004: 43).

The Indian scientific establishment has been working on the development of various renewable energy systems since the early 1980s. The MNES also gives financial assistance to established institutions as a part of its awareness programme. IREDA, a financial institution established in 1987, complements the programme Ministry through market development assistance and soft loans. IREDA has as of 31 December 2004 sanctioned loans of more than Rs. 6,667 crore and disbursed Rs. 3,620 crore. The Common Minimum Programme of the Government envisages electrification of all households by the year 2009. In addition, it is expected that the share of renewable power in the power generation would be around 10 per cent (MNES, 2005: 5).

Presently renewables contribute about 5,700 MW of power, which represents about 4.99 per cent of the total installed power-generating capacity from all sources as shown in Table 5.3. Wind power contributes about 2,980 MW, while biomass power and cogeneration account for 727 MW and the share of small hydro power 1,693 MW (see Table 5.3).

ONGC is expected to start wind power generation and producing bio-fuel as part of its efforts to harness alternate energy sources, in the next 2 years. Proposals in this regard have been submitted before the state governments of Rajasthan and Andhra Pradesh. Abdul Kalam expressed satisfaction at this development saying, 'The only way to meet (future) needs is to go the route of renewable energy apart from further exploration of deep sea oil resources and the enhancement of recovery factors' ('ONGC Eyes Alternate Energy Sources', 2005).

Table 5.3 New and renewable sources of energy (potential and cumulative achievement as on 31 December 2004)

	Potential	Cumulative achievement
Biogass plants	120 lakh	36.71 lakh
Improved chulahs (burners)	1,200 lakh	339 lakh
Wind	45,000 MW	2,980 MW
Small hydro	15,000 MW	1,693 MW
Biomass gasifiers		62 MW
Solar PV	20 MW/sq.km	191 MW*
Waste-to-energy	1,700 MW	46.50 MW
Solar water heating	1,400 lakh sq.m Collector Area	10.00 lakh sq.m Collector Area

*Of this 105 MW p SPV products have been exported.

Source: Annual Report, Ministry of Non-Conventional Energy Sources (MNES), 2004–2005, p. 2.

It needs to be emphasized here that the RE sector is not performing up to its potential. The RE sector is mostly state financed and there is little private investment. The lack of synergy between the RE sector and power sector and the regulatory commissions does no good either in this situation. The commercial support for research and development and the more meaningful and effective awareness programme are long overdue. Here NGOs, voluntary organizations and co-operatives need to be encouraged to play a more active role. Banks need to be encouraged to provide finances for RE-related projects. Lastly, the substandard quality of RE systems has given a bad name to the sector and for this overnight manufacturers need to be questioned.

India's search for energy in the 'extended neighbourhood'

India's interests in the Asia–Pacific can be better understood through India's 'extended neighbourhood' policy, encompassing its 'look east' policy. India's relations with South-East Asia have historical roots and current engagements with the Association for South East Asian Nations (ASEAN) and the Asia–Pacific follow these historical linkages.

To begin from the modern period, during the 1950s and 1960s India and the South-East Asian states shared common ideas in various forums such as the Group of 77 (G-77), Non-Aligned Movement, the United Nations and many economic forums as well. Since the Bandung Conference of 1955 India has striven for a closer relationship with South-East Asia.

In March 1993 India became a Sectoral Dialogue Partner of ASEAN and in December 1995, recognizing India's growing political, economic and strategic partnership with ASEAN, in Bangkok, the ASEAN Heads of Government made New Delhi a Full Dialogue Partner. The 1996 Indo–ASEAN conference cemented India–ASEAN relations. Subsequently, the Indo–ASEAN Joint

Cooperation Committee was established to provide substantive content to and implement programmes of co-operation in trade and investment, science and technology, tourism, infrastructure, human resource development, and people-to-people contact. In July 1996, Bangladesh, India, Myanmar, Sri Lanka, Thailand Economic Cooperation (BIMSTEC) built the bridge for economic co-operation between South and South-East Asia.

India's role in South-East Asia was endorsed by Liu Jianchao, the Chinese foreign ministry spokesman, saying,

> China and India are two important countries in Asia. We welcome the development of relations between India and the Association of Southeast Asian (ASEAN) countries … We hope that the relevant parties could promote efforts to maintain regional peace and development … India's ASEAN diplomacy will surely make further contribution to the social and economic and development of the region.
>
> ('China Hails India–ASEAN Summit', 2002)

India–ASEAN trade reached a high of US$7.35 billion in 1999–2000 and subsequently Indian exports to ASEAN grew by about 30 per cent (Ministry of External Affairs, 2002: 4). India enjoys strong relations with Brunei, Cambodia, Fiji (Pacific), Indonesia, Laos, Malaysia, Philippines, Thailand, and Vietnam.

Reaching out

Indian companies especially IOC, OIL and GAIL have been vigorously pushing their foreign-energy ventures as a consortium in the field of exploration, transportation and marketing of oil and gas. There are key gas producers in Indonesia, Malaysia, Myanmar and Thailand that interest not only India but also Japan, Korea, and China. Indian firms have acquired equity oil in Myanmar, Australia and Ivory Coast and are searching for other prospective options in the Asia–Pacific region.

OVL and PetroChina are pushing their energy agenda rigorously in the region but India has to further intensify its efforts to keep pace with the energy competition. Indian companies have secured a 30 per cent share in the Korean-led purchase of a gas field in Myanmar. ONGC operations cover oil and gas exploration, IOC covers marketing and GAIL covers transport. OVL is playing a leading role in acquiring oil equities overseas and has bagged some very lucrative deals abroad, which should encourage other Indian companies to join the league.

India–Myanmar

OVL's oil and gas ventures began in December 2002 and currently other areas are being explored for importing gas from Myanmar and also Bangladesh. The centrality and vastness of the Indian market in any major project is a vital factor, since India accounts for more than 44 per cent of the gas consumption in South

Asia. Mohuna Holdings, a Bangladeshi company, has also mooted the idea of laying a 'Trans–Myanmar–Bangladesh Gas Pipeline' within the BIMSTEC framework to carry gas to India (Muni & Pant, 2005: 149). If realized this would be a
significant breakthrough in India's efforts. India, Myanmar and Bangladesh
issued a trilateral press statement on 13 January 2005 on energy and power co-
operation, which indicated the existence of a nascent (particularly in
Bangladeshi) political will to harness the energy resources through a pipeline
from Myanmar to India through Bangladesh (Ministry of External Affairs,
2005b). Despite being a win-win proposition for all, the project remains hostage
to political and diplomatic vagaries especially between India and Bangladesh.

Myanmar has also hinted at an offer of more PSC oil and gas blocks, adjacent
to the A-1 block, to India. India–Myanmar energy co-operation has encouraging
prospects and India appears keen to venture in the hydro-power sector as well.
Initial studies are being undertaken into collaboration with Myanmar Electric
Power Enterprise for a hydro-power project on the Chindwin River at Hitamanthi
across India's north-east state of Nagaland. It is being estimated that 1,200 MW
would be produced for Myanmar's domestic consumption as well as for export to
India (Ministry of External Affairs, 2005b: 154).

India–Vietnam

The 1995 discoveries of large gas reserves in Vietnam are being keenly contested
between Korea, Japan, Thailand, Malaysia and India. Vietnam, the third largest
energy source in South-East Asia is ranked third in the region in hydro-power
potential after Myanmar and Indonesia and the second largest in coal reserves.
The handicap with Vietnam is the lack of refinery capacity for its crude oil. Its oil
reserves are located in offshore fields across the contentious Spratly Islands
located in Cuu Long, Man Con Son, Song Hong, Malay and Tho Chu Sea Basins
(Ministry of External Affairs, 2005b: 165–169). The four largest crude oil buyers
from Vietnam are the United States, Japan, China and Singapore and in 2002 a
5,000-km long pipeline was planned to link gas fields in Indonesia with Vietnam,
Malaysia, Thailand and China through the Asia–Pacific Economic Cooperation
(APEC) forum at the cost of $8 billion (Ministry of External Affairs, 2005b: 194).

India's energy co-operation with Vietnam germinated in the late 1980s during
the late Rajiv Gandhi's visit to Vietnam and under the framework of Vietnamese
Doi Moi (economic renovation policy). In that period India and the former Soviet
Union became aware of substantial oil and gas prospects in Vietnam. In 1991,
ONGC jointly with PetroVietnam, and British and Norwegian companies pitched
in to develop gas in the Nom Con Son basin. After some inter-company legal hurdles were removed, the project was eventually signed in 1999 between ONGC,
British Petroleum, Amoco and Statoil (Danish company) with PetroVietnam for
the production and transportation of gas, and power generation from Lan Tay and
Lan Do fields. The project entailed laying down a 370-km long pipeline from
fields to the shore at Ho Chi Minh city at a cost ranging between $1.5 and $507
million for the upstream segment. OVL accounted for 45 per cent of the upstream

project, followed by BP's 35 per cent and PetroVietnam's 20 per cent (Ministry of External Affairs, 2005b: 200). As per the official Indian figures, OVL invested $162 million in the project (Ministry of External Affairs, 2005a).

On May 1 2003, India and Vietnam also signed a 'Joint Declaration' on business management and consultancy activities in the fields of oil, gas, chemicals and petro-chemicals (Ministry of External Affairs, 2003a). Currently there are no specific Indian projects in Vietnam and the Nom Con Son project stands completed.

India–China

The two 'emerging economies' and most populated countries are projected to face a severe energy crunch by 2020, prompting them to forge a co-operative mechanism to face future energy exigencies. Despite being Kyoto signatories both are aware of their obligations to develop environmentally friendly sources of energy and are also cognizant of the perils of oil supply disruption due to unforeseen geopolitical factors. Both plan to complete their strategic oil reserves by 2010.

Analysts locate several common choices, imperatives and alternatives before the two countries. First, the need to devise a policy framework based on the current demand-supply trends to cater to long-term energy requirements; second, to invest in alternative energy resources such as renewables to reduce dependence on fossil fuels; third, to develop their domestic energy ventures and bring them to par with the big energy powers; and fourth, to tap overseas options and diversify energy import sources (Li, 2005: 3).

In 1997, India and China signed a memorandum on jointly exploiting oil reserves in developing countries, especially in the Sudan. In June 2003, India and China signed a Memorandum of Understanding (MoU) for Enhanced Cooperation in the field of Renewable Energy. The MoU seeks to:

> establish cooperation in the field of small hydropower, wind power and other areas of renewable energy through joint research and development activities, exchange of technical expertise and information networking. The ultimate objective is to commercialize the result of such cooperation, create business opportunities and facilitate sustainable market development in an environmentally responsible manner.
>
> (Ministry of External Affairs, 2003b)

During Chinese Premier Wen Jibao's visit to New Delhi in April 2005, the two sides agreed to step up their energy co-operation and India expressed its willingness to involve China in the Iran–Pakistan–India gas pipeline (Ministry of External Affairs, 2003b: 6). India keeps China posted on the developments in the pipeline, which may be extended up to China. India is keen to join China in acquiring oil and gas properties in third countries, which would convert competition into co-operation, to some extent, to mutual advantages.

India–Japan

Indo–Japanese relations, after the 1998 Indian nuclear tests, began to recover with Prime Minister Yoshiro Mori's visit to India in August 2000, the first time in 10 years by a Japanese premier to South Asia. In recent months India's and Japan's common concerns were further reinforced when they along with Brazil and Germany clubbed together as G-4 aspirants for the United Nations Security Council. As yet there is little activism between them in the energy field and they will have to work to realize their shared concerns and goals. Indian Defence Minister Pranab Mukherjee, in January 2005, at the Asian Security Conference organized by the Institute for Defence Studies and Analyses said, '… India and Japan have a convergence on energy issues and have joint concerns about the security of sea lines of communications and vital choke points in the Indian Ocean … ' (Raman, 2005).

In May 2005 Japanese Prime Minister Junichiro Koizumi visited India, which underlined the growing desire on both sides to strengthen bilateral relations in every sphere. On 29 April 2005, Indian Foreign Minister Natwar Singh, at the inauguration of the India–Japan Parliamentary Forum in New Delhi, observed,

> The launch of the India–Japan Forum at a time when Prime Minister Koizumi is in India, sends out a clear signal about the importance we attach to India–Japan relations at all levels of interaction … at the political level the two countries must further fortify their cooperation and pursue an all round and comprehensive development of bilateral relations … and fully utilize their dialogue mechanisms as also establish newer ones in areas considered necessary, such as energy security.
>
> (Singh, 2005)

The subsequent Joint Statement issued among other areas also touched upon energy security. It says that the:

> [t]he two Governments will also strengthen their energy and environmental cooperation, including on sustainable development and environmentally sound technologies. In doing so, they will ensure increased focus on energy security, energy efficiency, conservation, and pollution-free fuels. They will also cooperate in the hydrocarbons sector in areas such as exploration and production, and downstream projects including in third countries as well as on improving Asian oil markets and increasing investments in Asian energy infrastructure.
>
> (Ministry of External Affairs, 2005c)

Both India and Japan are energy-importing countries and need to forge co-operative measures to meet with their future energy requirements, however, any substantial result on the ground is yet to be seen. Presently traditional security and trade issues are dominating bilateral engagements. India is vital for the security of the choke point in the Malacca Straits against piracy, a point which both China and Japan acknowledge.

Security of Sea Lines of Communications (SLOCs)

Maritime trade, originating in the Persian Gulf, travels through several vulnerable points before reaching the South China Sea, East Asia and the Asia–Pacific, and is always susceptible to piracy and terrorist attacks. Being the preferred route for 40 per cent of the world's oil trade, this route is referred to as the 'New Silk Route', passing through the Strait of Hormuz in the west and Malacca and Singapore in the east, linking the Persian Gulf to the Arabian Sea and Indian Ocean. Around 15 mmb/d of oil crosses this choke point, including around 60 per cent of imports to China and India. Through the Malacca–Singapore Strait more than 62,000 vessels pass each year (Khurana, 2005: 6).

The oil vessels are difficult to manoeuvre and with a LNG cargo they are environmentally hazardous as well. The Malacca Strait is only 1.2 nautical lines wide at its narrowest and 22 miles deep at the shallowest. Over the years due to intense fishing, shipwrecks and shoals they pose a serious pilotage challenge to vessels. This was accentuated in the wake of Tsunami in December 2004 which has reconfigured the underwater topography (Khurana, 2005: 9).

Numerous cases of drug and arms smuggling and piracy have been reported in the past, which have threatened to upset the fragile maritime security apparatus of the region. The year 2000 witnessed a rise of 56 per cent in piracy incidents from 1991. In the Malacca Straits, 75 cases of piracy were reported in 2000 as against 32 in 1991. In between, the numbers had reduced drastically, and in 1997 no such cases were reported. In 1998, there was just one piracy incident. Bangladesh reported 55, India 35, Indonesia 119, Malaysia 21, Myanmar 5, Philippines 9, Singapore Straits 5, and Thailand 8 cases of piracy and shipping robbery cases in 2000 (Abhyankar, 2001: 23–24).

The security of the SLOCs is a complex affair due to the territorial sensitivities of the littoral countries and therefore makes their endorsement imperative for any security mechanism envisaged. Geoffrey Till considers the Malacca Straits as an area where piracy and illegal trafficking of all sorts could endanger the peace and security of the littoral countries as well as inflict considerable harm to the interests of those with ties to the region. However, he cautions that any naval co-operation or anti-piracy measures have to keep the political sensitivities of the littoral or local states in mind (Till, 2001: 51–52).

India always factors the concerns of the littoral states into its maritime security policy. India's growing bilateral relations in the last decade or so with countries like Singapore, Malaysia, and Indonesia make it more suitable for the task of ensuring SLOCs' security than China and Japan. India's role in ensuring smooth energy supplies through the Malacca Straits could be considered crucial for the energy concerns of countries in South China Sea, East Asia and the Asia–Pacific.

To strengthen security mechanisms in the Malacca Straits, the Indian and the Chinese navies could devise a co-operative mechanism against piracy and terrorist attacks. In this regard, the Regional Cooperation Agreement on Combating Piracy against ships in Asia is a welcome measure which would ensure safe imports to India as well.

Conclusion

In securing the energy future for its coming generations, India has joined the global frenzy in exploring all possible domestic and external avenues. Over a decade or so, the Indian government has relentlessly endeavoured to strike a balance between its domestic and foreign-energy policies, through necessary changes in its existing energy policies. India well understands domestic reserves will not suffice to meet its future demands and the wedge between supply and demand will widen further by 2015 or 2020. This has triggered a vigorous energy hunt by India though not with the same degree of successes as China, Japan have achieved. India's geo-strategic location, growing industrialization, booming population, need for better technologies and growing costs in the energy sector make the task highly challenging. Being an emerging economy India has to steer its policies and search for energy resources within its financial and technological constraints, while enhancing its capacities on both counts simultaneously.

Heavy dependence on oil imports, the global hike in crude oil prices and growing geopolitical uncertainties in the Middle East and the Gulf region do not paint a promising picture for India. In this context, India's efforts at diversifying its energy supplies from overseas locations other than the Middle East and the Gulf are timely and sensible. In addition, given the enormous costs involved in securing energy from abroad, the introduction of the NELP is a prudent move which has enabled the private sector to further technological, financial and logistical support. The synergy between ONGC, OIL, and GAIL in exploration, processing and transportation has brought encouraging results and therefore carries a fair bit of promise as far as India's energy future is concerned. The government has been equally up to the mark by envisaging the Hydrocarbon Vision 2025, NELP and the X Plan which seek to harness domestic and foreign energy resources with increased private sector involvement. The plan to complete the Strategic Storage by 2010 is another facet of India's long-term thinking. On the domestic front, the successes of private companies such as Reliance Industries in the KG Basin have instilled vigour to the efforts of other private sector undertakings.

The MPNG and the Indian External Affairs Ministry have reached out to Bangladesh, Iran, Myanmar, Pakistan, and Vietnam among others, in expediting the possibilities of importing gas through pipelines. Geopolitical constraints especially in laying the 'peace pipeline' from Iran to India, through Pakistan are immense but the two ministries have faired well in overcoming them. A cost-effective option, which the 'peace pipeline' is an instance of what is becoming the preferred means for transportation of oil and gas worldwide.

For a sustainable and secure energy future the key is to devise a comprehensive strategy which encourages new policy formulations, private participation in the energy sector, establishment of regulatory bodies to oversee pricing and conservation efforts, and synergizes between domestic and overseas efforts. Political will is needed to introduce institutional changes and overcome geopolitical impediments in materializing pipeline dreams.

India's 'Look East' policy based on the 'extended neighbourhood' concept has enabled it to make inroads into South-East Asia and the Asia–Pacific. India's successes in Vietnam and near success in Myanmar widen the scope for such co-operative endeavours. The world is becoming more and more inter-dependent in a market-driven age and the challenges of traditional and non-traditional security have to be addressed collectively. India's strategic location places it in a unique position to help provide security for maritime trade to and from the South China Sea, South-East Asia and the Asia–Pacific. This could be India's contribution to the region's energy security efforts and could help secure its own imports as and when they begin from Vietnam. India's eastward energy search has so far brought dividends in Vietnam and to a large extent in Myanmar. In Bangladesh, there are still doubts being raised on the feasibility of the pipeline to India due to domestic political factors, but in principle it has found support. Indian companies can help countries in the Asia–Pacific such as Singapore and Vietnam in downstream projects and in turn negotiate for their share of energy.

Energy demands in India and China will double by 2030, a major challenge for both. The search for hydrocarbon reserves has to be supplemented by intensifying the search for renewable energy resources through harnessing of hydro, solar and wind power. This requires more investment, research and technological improvements. Besides, enhancing public awareness for saving energy through increased use of renewables has to be made part of the energy philosophy of the government wherein the role of NGOs and voluntary organizations becomes pertinent. Sharp rises in vehicular transport raising pollution levels makes it necessary to gradually switch liquid fuel to gas, which would be cost effective and environmentally friendly. Research findings that more gas is available than oil in the world present the perfect setting for the switchover.

The responsibility to secure the energy future hinges on people's participation as well and they have to play their part in promoting efficient uses and conservation of energy. Very few use solar cooking methods to conserve energy and there is a need to adopt such means for other purposes such as boiling water and courtyard and backyard lighting. In cities, due to rising prices of petrol, people have worked out car pooling systems which help in saving fuel, reducing vehicular density on the roads and cutting down individual fuel expenses. The government's endeavours will show visible results in about 10 years and similarly peoples' contribution no matter how minimal it may seem now, would make a difference in the long run.

Part II
Supply

6 Australia and Asia–Pacific energy security: the rhymes of history

Richard Leaver

History does not repeat itself, but it does tend to rhyme.

(Krugman, 1999)

... all great world-historic facts and personages appear, so to speak, twice ... the first time as tragedy, the second time as farce.

(Marx, 1852)

In these post-communist times, this is about the only bit of Marx that is remembered. Unfortunately, on the topic of energy security – a field constituted whenever energy policy either falls or is pushed into the orbit of national security – Marx did not go far enough. His characterisation addressed only the second turn of the historical wheel, whereas we live today at the beginning of a third age of heightened awareness about energy security. It began in 2001 when energy security was installed as an official objective in China's tenth five-year plan, and it is perhaps good news that the country that presently accounts for the lion's share of the increase in global oil demand has embraced that idea. On the other hand, the need for a third appearance also suggests that not enough was achieved during earlier comings. The spectre of repeated under-achievement therefore begs the question of what comes next in Marx's descending sequence from tragedy to farce. The musical?

A broad-brush outline of previous ages of energy security in the post-war period ages is needed. The original conjunction of energy policy with national security was confined to the United States – but at a time when its powers were at their height, and where the extra-territorial effect of US domestic policy was therefore extensive. In 1959, the Eisenhower administration invoked national security to strictly limit the extent of US dependence on imported oil to just 9 per cent of anticipated future consumption. This was the period when the massive discoveries made in the Middle East over the previous two decades were just beginning to enter world markets in significant volumes, with US-based oil majors doing most of that marketing. It was also the moment when the US political position across the Arab world was considerably enhanced by its role in bringing the Suez War to a halt, and when the US share of world oil consumption was well over 40 per cent. Given the combination of all these commanding positions, theory tells us that the US hegemon should have promoted an international free market in oil. Instead it went the other way towards 'energy independence' – while simultaneously encouraging other oil-importing industrial countries to embrace higher dependence on Middle Eastern oil. The malignancies established by this divergence became so pronounced that polite people soon chose not to recall this first experiment in energy security (Leaver, 2004).

With memories suitably purged, the field of energy security then appeared to commence with the 1973 oil shock. And to a post-war world accustomed to ever-cheaper oil, the quadrupling of prices that commenced in the Yom Kippur war was indeed a considerable watershed.[1] When Organisation for Economic Co-operation and Development (OECD) countries reacted to the novel exercise of 'producer power' by creating a united front of consumers through the International Energy Agency (IEA), energy security through co-operative regime creation promptly became an institutionalised part of the diplomatic landscape. IEA member states encouraged themselves to establish national oil stockpiles that would provide insulation against future shocks; to develop protocols for sharing their stockpiles in another supply crisis; to diversify out of oil where feasible; and to engage in various demand-side forms of energy conservation. When, however, oil prices fell dramatically in 1986, many of the co-operative elements from the second age of energy security began to seem like expensive indulgences. This time, many chapters in the IEA manual were quickly forgotten.

If we are indeed at the dawn of a third age of energy security, then it is notable that Australia's relationship to this new age is unique and seems to offer co-operative policy possibilities, at least on a regional scale. For in spite of a small deficit in the liquid fuels balance, Australia has for some decades been the world's largest exporter of coal, and the largest by a considerable margin. In uranium, its long-known dominance of low-cost reserves looks as though it could soon translate into a leading position in current production. Likewise, although Australia presently exports less liquefied natural gas (LNG) than Indonesia or Malaysia, its gas reserves are already the largest in the region, and may soon exercise a decisive influence over the rank order of regional producers. It is unlikely, furthermore, that domestic energy requirements, although buoyant, will diminish any of these export profiles: Australia consumes no uranium at all, arguably too little natural gas, and buckets of coal – but not, primarily, the higher hard grades that go to exports. Hence, in a region where the general trend is clearly towards lower levels of energy self-sufficiency, Australia is already a net energy exporter of considerable significance – and the degree of that significance looks likely to increase. Australia appears to have the wherewithal that will allow it to enter into regionally scaled co-operative ventures in the name of energy security.

But this was equally true in the second age – and yet Australian governments of both persuasions 'defected' from the IEA framework for the first 4 critical years of its existence. Hence, at the beginning of the third age of energy security, with real oil prices already substantially higher than they were during the period of Australia's self-imposed IEA alienation, those with a sense of history's rhymes should look back with suspicion when contemplating the co-operative possibilities arising from Australian energy policy.

French high-tech dreaming

When Australia sat on the IEA sidelines, it had company on the bench. Although the OECD offshoot was based in Paris, France did not become a member until 1992. This odd couple was doubly odd in the sense that the impact of the 1973 oil crisis upon them had been vastly different. Like the rest of Western Europe, France imported all of its oil, and mostly from the Middle East. But its domestic coal industry had declined particularly sharply through the Long Boom, and France was therefore somewhat like Japan in that oil's share of total primary energy was extraordinarily high. In these extreme circumstances, the 1973 shock had a dramatic impact upon both the balance of payments and the character of economic growth in France. On the other hand, Australia's oil import dependency stood in 1973 at just over 30 per cent – virtually identical to the relatively privileged US end of the spectrum of vulnerabilities. Since oil was generally not used in electricity utilities, its share in total primary energy was relatively low. And furthermore, the price paid by refineries for Australian crude had been administratively fixed at $2.06 per barrel in 1970 – and was destined to remain at this level for some years to come. So although the Australian economy, like that of every other OECD member, ground to a halt after the oil shock, it was indirect causes stemming from the collapsing world economy rather than the direct cause of rising petroleum prices that played the lead role in the local production of that drama.

In spite of these vast differences, what united Australia and France was the belief that their control of other kinds of non-oil energy resources could provide them with significant political and economic opportunities in rampant hydrocarbon markets. For each of them, the strategy of going it alone as a vendor of energy and energy-related technologies had more short-term attractions than OECD-scale consumer co-operation. For the French, their perception of unilateral advantage was bound up with the leverage that nuclear technology might provide to French energy diplomacy. For Australia, the advantage was expected to come primarily from the low end of the spectrum of techno-optimisms.

France already had a significant pre-history of using nuclear exports in the service of foreign policy, having supplied Israel with the Dimona reactor in the late 1950s.[2] But it had also recently obtained a new degree of freedom by virtue of its non-membership of the NPT when, in 1970, that treaty swung into operation. Since it was a recognised nuclear weapon state with a complete nuclear fuel cycle, France was initially at the outside edge of the NPT's cobweb of 'gentlemanly agreements' that sought to impose some degree of supply-side control over nuclear exports. And in 1973, France drew upon both its active tradition of nuclear export diplomacy and its new-found freedom from NPT restraint to commence negotiations with Pakistan that would eventually transfer most of the technology needed to reprocess spent nuclear fuel.[3] It followed this up in 1974 by giving Iraq, one of France's largest oil suppliers, a leg up into the nuclear age. France's end of the Iraqi bargain entailed a replica of their Saclay reactor, a large capacity research reactor whose most unusual characteristic was its fuel load of weapons-grade high enriched uranium (HEU). Although used in France for testing nuclear materials, it also had the capacity to produce significant amounts of plutonium.

This faustian 'oil-for-technology swap' proved over time to be pretty much the last gasp of France's nuclear diplomacy. Iraq's share of France's national oil imports did rise, but only by a disappointing 6 per cent (although not insignificant advantages were obtained through Iraqi purchases of more conventional French military assets). But the purposes to which the Iraqis intended to put their Tammuz reactor progressively became more suspicious, even to the French.[4] Then, in mid-1981, apparently only a matter of weeks before being commissioned, the reactor was destroyed by the Israeli Air Force.[5] By this stage the Iran–Iraq war had already brought about the dramatic collapse of Iraq's oil exports. So although Paris stood rhetorically alongside Iraq's attempts to rebuild the reactor, it was notable that they approached this task with much greater scepticism than was previously evident.[6] The finishing touches were applied immediately after the 1991 Gulf War had revealed the full extent of Iraq's renewed bomb programme, for France then occupied its seat at the high NPT table of recognised nuclear weapon states that had been kept warm for it for two decades – and ramped its nuclear safeguards policy up to the full-scope standard that this occupancy required.[7] By that time, paradoxically, nuclear technology had effected a rather different solution to France's 1973 energy dilemma, with 75 per cent of all French electricity coming from nuclear reactors. Any return to the earlier practice of carefree nuclear exports became hard to imagine.

Australia: the opportunity within the crisis

During the period of France's alienation from the IEA, Australian governments also had their own version of nuclear dreaming. Characteristically, however, it mainly centred upon the raw materials of the nuclear fuel cycle rather than its most advanced technologies.[8] For if France's oil-for-technology swaps anchored the high-tech end of the IEA rejectionist front, then a rather loose Australian notion of 'resources diplomacy' tied down the low-tech end.

Since no minister in any Australian government ever celebrated this term, attempts at exegesis therefore had to rest upon the very occasional speech by one or another senior bureaucrat (Arndt, 1974). In this official vacuum, analysts tended to sheet home the whole idea to Labour's obtuse and intensely nationalistic Minister for Resources, Rex Connor, who had indeed arrived in office with concerns about the low prices achieved on export markets by Australian minerals and with plans to impose controls if they did not improve. But the impetus to resource nationalism was more widely spread around, politically speaking. It went back at least to the Gorton government's worries about the excessive level of foreign investment in the domestic mining industry – a concern rather romantically referred to as 'buying back the farm'. And even though Gorton's successor, William McMahon, was the first great Australian believer in free-market virtues, he nonetheless allowed important exceptions when dealing with the minerals sector. It was his government, for instance, that aided and abetted the construction of a uranium cartel centred on a process of sequential bidding in the international yellowcake market – a practice not exposed until its outing in the Mary Kathleen papers 5 years later (Stewart, 1981). By this stage, talk of resources diplomacy in Australia was transforming itself into something closer to NIEO thinking, a liberalised variant of petro-power where the reform of 'the rules of the game' in international trade was thought preferable to further revolutions at the hands of producer associations (Barclay, 1978).

With no obvious intellectual parent or clearly articulated stratagems, Canberra's resources diplomacy rested primarily on two domestic resources booms that pre- and post-dated the 1973 oil crisis. The first of these booms took place in the late 1960s and early 1970s, when a range of new mineral deposits in both the fuel and non-fuel categories were discovered (Murray, 1972; Sykes, 1978). The significance of these discoveries was then amplified in successive stages. The first amplification came from a wave of popular interest in the stock market, which unleashed the characteristic mass psychology of unbounded opti-mism that, in turn, sowed the seeds for the 1971 market collapse. But there was much more than just speculative froth and bubble to this boom, for the second stage of amplification had more material roots. This, after all, was the period when minerals became the champions of Australia's balance of payments. And leading the new export order were two industries intimately linked as input suppliers to Japan's emerging world-class steel producers – namely, iron ore and coking coal. And as public finance was not central to the expansion of either of these industries, they were able to ride out the stock market gyrations that shook most other miner-als (Byrnes, 1994: 68–71). Indeed, more or less simultaneous with the deflation of Australia's resource bubble, Japan seized the title of the world's largest steel pro-ducer, and the future of Australian minerals exports seemed assured.

But when the OPEC revolution quadrupled the price of the world's backstop energy, a second Australian boom got under way. This was like the first boom in having two boost phases, but unlike it in that it was narrow-cast upon the energy sector. The first boost came from export promotion effects, and saw a new Australian export market for steaming coal created as Japanese electricity utili-ties rapidly abandoned the use of fuel oil to power their generating sets.[9] The sec-ond boost came from import substitution effects. With domestic power generation rooted in the lower grades of coal and controlled by state governments that were willing to ignore environmental externalities, domestic Australian energy prices moved towards the bottom end of world markets. Australia thereby became one of the preferred sites for energy-intensive global industries such as aluminium smelting, an emerging regional industry whose centre of gravity had previously been located in Japan.[10] Aided by the Japanese bureaucracy that was eager to minimise their nation's energy import bill, Australia was able to upgrade its position from the world's largest exporter of the raw material alumina to a globally significant exporter of refined aluminium in little more than a decade (McKern and Waltho, 1988: 108–115).

Spanning the two booms, and to some extent uniting them, was the renewal of interest in uranium. In 1970, after uranium exports to the British–American bomb programmes had run their course, new discoveries of high-grade uranium ore were made in the Northern Territory. And half a decade later, when the Olympic Dam deposit was found in South Australia, Canberra suddenly found itself in possession of at least 30 per cent of world reserves of low-cost uranium. Although there was a long pre-history of official anticipations of a civil-oriented uranium export industry, its realisation had come to be immensely complicated by the growth of mass-level anti-nuclear sentiment.[11] Eventually, under the crafty guidance of the Fraser government, a combination of inflated export earnings and self-projections of Australia's diplomatic importance to the integrity of the civil nuclear industry opened a small chink in the wall of domestic opposition through which the uranium export industry could and did expand.

Coming on board

There still remains a considerable mystery about exactly what it was in the makeup of the IEA that annoyed two quite different Australian governments for 4 years. According to the considerably opaque diplo-speak of the Agency's official history, Australia had concerns 'about its responsibility to provide petroleum products to nearby countries, about cooperation in bunkering for shipping or for particular grades of oil and about cooperation between producers and consumers in international commodity trade' (Scott, 1995: 52). But had Australia become an early member, it might well have argued that the steps it was taking to develop and export other energy forms were entirely consistent with the IEA's purpose of diminishing the possibilities for further oil shocks. Since, on the backside of the second oil shock, the IEA did come to accept that structural change in the composition of primary energy was a frontline concern, early Australian membership may well have hastened its movement along this policy pathway.

The most likely cause, therefore, of Australia's alienation lay in the IEA's somewhat obsessive initial focus upon the development of protocols for the emergency sharing of oil in circumstances where the supply to any consumer state fell by 7 per cent. In these crisis circumstances, member states were required to accept what could be depicted as quite extraordinary derogations of sovereignty to the IEA secretariat over the sharing of oil, derogations that could only be overridden by a special majority of members. One must presume that Australian governments, being relatively self-sufficient, saw no pressing need either to build up above-ground stockpiles of oil or to share those expensive indulgences with other countries which were primarily consumers rather than producers of oil – and doubly so when this critical standpoint difference would place Canberra in a permanent minority inside the organization. As a producer and net exporter of energy, the chance to passively profit from the age of expensive energy made the alternative to possibly costly IEA membership seem quite attractive.

The problem with this line of interest-based arguments is that much the same reasoning applied with equal vigour to the United States – and yet this was the country that had done most to bring the IEA into existence and breathe policy life into it. Australia's self-imposed alienation therefore entailed the widening of differences with Washington, something that no Australian government likes to do without good reason. It was notable, therefore, that the 1979 Australian accession to the IEA came on the backside of two critical shifts in US policy, one outside and the other inside the IEA.

The exterior shift was evidenced in the shape of Carter's 1978 Nuclear Non-Proliferation Act, where the US sought to articulate a coherent alternative to the European and Japanese drift towards 'the plutonium economy'. Instead of passively watching the slow geographical spread of advanced stages of the nuclear fuel cycle, US policy sought to encourage allies to expand along the more proliferation-proof pathway of the light water reactors, low enriched uranium fuel and the once-through fuel cycle. And Australia, along with Canada, had important roles to play in Washington's new script as uranium suppliers. For if reprocessing were to be discouraged, then demand for new uranium would have to increase quite dramatically. But civil nuclear customers could not be expected to follow this American pathway if they were in the least bit uncertain about the future of uranium supply; hence the importance of the Australian debates about the resumption of uranium exports.[12] Hence, also, the importance of heightened IEA interest in new fuels in general – uranium obviously, but also coal – that was easy listening to Canberra's ear. And indeed, only 2 months after Australia's accession, the IEA's governing board agreed upon a package of *Principles for IEA Action on Coal* that sought to expand coal trade by facilitating commercial negotiations between coal-consuming and coal-producing nations. Australia's interests as a non-oil energy exporter were largely reconciled with the purposes of the IEA, ostensibly a united front for consumers.

But the Agency's purposes were in any case shifting in the face of a big American push towards the demand side of the oil market. Up to 1978, Washington had imposed no restraints at all upon its own imports of oil. Driven along by the combination of administratively imposed low petrol prices and the relative success of Keynesian stimulation in returning the economy to the path of economic growth, the share of oil imports in final consumption had soared by nearly 20 per cent (Kahler, 1980). In 1978, however, Carter capped US oil imports at the previous year's levels – and sought to establish import targets for other IEA countries. Beginning more or less from the point of Australia's accession, there was a brief period of strong US backing for a targeting policy in a variety of formats (Keohane, 1982: 471–474). Initially collective in form, the IEA sought to bring down the aggregate oil demand of agency members by two million barrels per day. It rapidly became clear, however, that the US wanted national targets – and that most members would only commit to them if they were allowed to contain a generous measure of free play. Soon, however, the coming of a second oil shock and its associated recession were shrinking national oil imports everywhere, and the contentious US policy of consciously imposed targets was becoming superfluous. It was relegated to the category of desperate measures, the residency to which the new shock had already consigned the trigger policy of oil sharing. With triggering and targeting off stage, IEA policy shifted towards areas that Canberra found much more congenial – supply-side structural change of primary energy forms in particular, tweaked by the uncertain moral force of periodic reviews of the energy programmes of Agency members.

Problems and profiles

Like France on the high-tech road, the Australian governments that travelled the low-tech road of resources diplomacy also ran into problems. The results were, at best, a mixed bag. Uranium, in particular, never lived up to its advertised potential. This should have been no great surprise, since the industry had been rather shamelessly inflated at the highest political level in the effort to buy the necessary minimum of public support. So whereas the 1977 Ranger Report had cautiously suggested export revenues of A$500 million (in 1976 prices) by 1984–1985, rising thereafter to A$1200 million by the beginning of the 1990s, the Fraser government was always in the hunt for something bigger. Deputy Prime Minister Doug Anthony eventually settled on Donald Barnett's bullish 1979 projection of A$1,800 million per annum – a calculation that required the heroic assumption that prices would remain high (at A$90 per kilogram) while Australian production soared (to 20,000 tonnes) (Barnett, 1979: 156). Had these heroics been capable of reconciliation, then uranium exports would have been more than sufficient to close Australia's expanding trade gap. In the event, long after Anthony's retirement, the actual volumes and returns for uranium exports were still coming in well under the Ranger estimates.

With coal, however, the Australian problem came from over- rather than under-performance. At the level of national aggregates, the level where uranium was failing, there was never an issue with coal. Steaming coal volumes, initially not even 10 per cent of coking coal, quickly leapfrogged over it, putting Australia on the road to becoming the primary global exporter of coal. This status was achieved in 1985, by which stage the four-fold increase in volumes over the previous decade regularly established coal as Australia's largest export. Another decade would see another doubling of volumes. But while all this volume expansion was going on, international prices generally trended downwards, quite spectacularly so after the 1986 oil price collapse. And falling prices plus rising volumes meant problems at the micro-level of the firm and the mining community – so much so that the coal industry would later be held up as a prime example of 'the profitless growth phenomenon' (McIntosh–Baring, 1993).

Not surprisingly, having never achieved any clear definition, the resources diplomacy idea simply faded away without much trace somewhere in the mid-1980s, with Treasurer Keating's infamous warning about declining terms of trade and the banana republic syndrome delivering a stake through the heart of the minerals export cult. Manufactures, especially elaborately transformed ones, were rapidly becoming the new Australian *zeitgeist*. But the Australian contribution to regional energy security was nonetheless impressive, although no government quite developed the voice to publicly celebrate it. By the time Keating was driving his stake home, Australia already stood in the company of the two North Sea oil exporters in terms of the energy equivalence of coal exports alone. When LNG exports from the North-West Shelf started up at the end of that decade, increased coal volumes, plus the less spectacular quantities of uranium, placed Australia quite clearly at the top of the short OECD list of net energy exporters. If there was little fanfare, it was largely due to the across-the-board decline of interest in the ideas of energy policy and energy security: both, it seemed, could be reduced to what Keating once called (in a different context) 'a beautiful set of numbers'. And after 1986, no number was more beautiful than the headline price of oil.

Indeed, ideas about 'free market triumphalism' that would achieve much great fame after the end of the Cold War had their roots in the beginning of the end of expensive energy – which in itself played a massive role (and a massively neglected role) in the political demise of the oil export-dependent Soviet Union (Morse, 1983). And if international communism could not survive the decline of oil prices, then it is small wonder that the weakly developed Australian conceptions of energy policy and energy security could also not be sustained.

Australia's third age dilemmas: the crisis within the opportunity?

Fast forward the better part of 20 years to the beginning of the third age of energy security. Ever since Kent Calder rather accurately foresaw its general outlines, it is an age where demand-side dynamics emanating from the Asia–Pacific have figured prominently (Calder, 1996: Chapter 3).[13] And one index of the rise of the Asia–Pacific is that the region's problem has become a global problem: earlier ages of high oil prices, so the argument goes, were driven by supply shocks, whereas the market is currently said to be driven by demand increases. And Chinese demand increases in particular seem to grind on in virtually all circumstances – including some very unlikely ones.[14]

But even with China included, today's global demand increases are so far below the annual 8 per cent per annum averages that were turned in throughout the Long Boom that the motif for energy security's third age has to be 'market failure'. The motif arises, as it inevitably must, out of oil market dynamics, where free-market principles were progressively alleged to triumph between 1983 and 1986. The 15 years of low oil prices that followed OPEC's first-ever posted price reduction in 1983 signalled repeatedly and spectacularly that the time was not yet ripe for new investment in refining or exploration in particular. Those signals were still being transmitted at the end of 1998, when oil prices reached pre-1973 levels after the collapse of Asian demand during the Financial Crisis. So although there are many contributory causes of today's tight oil market, the two most proximal are the choke points that developed during those 15 years in refining and exploration, both of which now rival the geo-strategic choke points of earlier times as standing sources of upward price stickiness. This market failure might, of course, have been expected if modern analysts had been more attentive to the periodic discussions of the (ir)relevance of free-market theory in the historical and institutionalist literature on oil (see Frankel, 1983); as Karl Popper argued, things that can be anticipated can always be acted upon and pre-empted. Unless, that is, it is a point of principle that the market, like the legendary customer, is never wrong.

But Australia, as previously, is very much the Evans-style 'odd man in' so far as the regional pattern of this third age is concerned. As conventionally understood in terms of the potential for supply disruptions, Australia's energy security hardly seems to be a pressing national problem, especially when compared to the acute import predicaments faced by many regional neighbours. And for a government that likes market mechanisms and dislikes vision – especially in foreign affairs[15] – the good news in the coming of energy security's third age is that it appears to require none. There presently seems to be possibilities for a legendary win-win market-driven outcome from Australia's energy dealings with the region. Much like the king tide of APEC thinking in Australia, there is a neat complementarity between the strong Australian export positions in coal, uranium and LNG and their reverse video energy deficits in the profiles of most East Asian economies. With energy prices rising across the board, all that seems necessary in the way of government vision is for Canberra to clear away whatever natural or unnatural domestic obstacles there are to enhanced supply – and then stand aside while the hidden hand of the market works its usual effect. The third age is, it seems, destined to be nothing more than the second age revisited.

Market failure, however, threatens to ruin the party by exposing a series of crises that lurk inside this opportunity. The balance of this chapter therefore looks at the Australian manifestations of market failure in each of the energy markets that are important to Australia – beginning with oil, progressing through uranium to LNG, and ending with coal.

Oil: scarcity looming?

Right at the end of energy security's second age, one prestigiously published analysis correctly observed that liquid fuel policy had been pretty much the sum total of energy policy in Australia (Marks, 1986: 72). And oil, of course, tended to bulk large for the simple reason that Australia was still a net oil importer. Indeed, governments seemed to understand just how lucky they had been in the historical timing of Australia's modest oil discoveries; barely 5 years before the 1973 oil shock, national self-sufficiency only stood around 10 per cent (Saddler & Ulph, 1980: 72). Hence liquid fuels had provided the perennials to the arena of public Australian debate about energy policy: about tensions with the IEA over the level of national stockpiles; about the poorly disguised official disinterest in demand-side measures; about the future trajectory of liquid fuel self-sufficiency; and so on.

By contrast, the Howard government has progressively become less obsessed with any of these usual suspects. This became obvious during the run-up to the

2001 federal election, when it de-linked the course of the liquid fuels excise from the consumer price index and made a one-off reduction in the rate at which the excise was collected. This populist ploy was highly successful in aligning regional votes behind the National Party, but its effects in other respects were considerably less positive. Quite apart from the damage done to the long-term structure of fiscal policy, the lowering of the forward trajectory of domestic petrol prices at a time when international prices were already rising propped up the less fuel-efficient end of the auto market. All this was done at a point in time when the peak of Bass Strait production more or less passed, with new additions to national reserves falling well short of current consumption (Campbell, 2003: Appendix II).

In the year immediately following this de-linking, Geoscience Australia began articulating some disquieting themes: about the fact that Australia's share of world production was twice as big as its share of world reserves, and about the sharp fall expected in the ratio between predicted actual production and consumption of crude and condensate from 89 per cent to 45 per cent during the next half decade (Geosciences Australia, 2002: 29–32). Although the government later enhanced the incentives for exploration, it seemed equally clear that it did not really care one way or the other about the trend to depletion, arguing in its June 2004 White Paper on Energy Policy that energy exports in other sectors could pay for increased imports of Middle Eastern oil (Australian Government, 2004). But when that same Paper attached itself to a forward oil price of US\$35 per barrel, the lucky country seemed to be morphing into a problem gambler (Fels & Brenchley, 2005).

Worse still, evidence from the auto sector – the sector most intimately implicated in the primary end-use for liquid fuels – soon suggested that the problem gambler was running into a losing streak. As international oil prices passed through the White Paper's presumed ceiling, the short-term effects of the 2001 excise reductions upon the composition of auto demand were progressively overwhelmed. Although SUV sales had increased immediately after those excise reductions, more recent evidence unambiguously suggests that the auto market, although still buoyant, is increasingly being driven by the demand for small cars. But small cars, which now account for 70 per cent of all new sales, are entirely imported, while domestic auto production is centred upon the relatively large 'Australian six' that is targeted at fleet managers (where sales tax exemptions prop up the front line of domestic demand). The 2005 surge in petrol prices has revealed that the General Motors' Commodore, the largest selling of the locally made cars, has been suffering declining sales ever since 2002, and that the temporary surge in SUV sales that followed the excise reduction has been thrown into reverse, with sales of larger units particularly affected (Scott, 2005). The August 2005 announcement that GMH would be closing down its third shift after barely 2 years in operation suggests serious counter-purposes between government signals and consumer preferences.

There are, of course, much larger causes of the continued loss of manufacturing jobs in the auto sector – and largest of all, the rise of China as a supplier of components to the local industry (which, in other markets, is already turning out to be the vanguard to the fully assembled Chinese-made car). Nonetheless, the audacity of the government's 2001 gamble with tax signals along the interface between the drivers of the supply and demand for liquid fuels became ever more striking with the passage of time. For with the help of some leverage from the auto sector, what that gamble has managed to expose is a modest energy security problem in liquid fuels that has deep social implications – implications broodingly hinted at in the recent remarks of the industry minister that the Australian auto industry has now reached a 'low water mark' below which it cannot survive (Gordon, 2005). In the face of this complex of market failures, the Howard government may finally be discovering some of the reasons why the avoidance of this kind of liquid fuels debacle was previously such a government preoccupation.

Uranium: a triumph of hope over reality?

The re-entry of Australian uranium into world markets in the early 1980s more or less coincided with the movement of that market into the phase that Thomas Neff has recently called 'the inventory liquidation era' (Neff, 2005). At first, this saw US government stockpiles of uranium built up for strategic purposes during earlier times being sold off. The process then moved into higher gears with the end of the Cold War, when concerns about 'the loose nukes problem' in the former Soviet Union gave a compelling reason for bringing the nuclear raw materials of the new post-Soviet republics to market. The Russians arrived in 1993 bearing that most special kind of inventory known as warheads, and in what was described as 'one of the more intelligent national security initiatives in US history', the US Department of Energy purchased and commercially disposed of Russian-origin reactor-grade fuel produced by blending down about 30 tons per annum of HEU (Falkenrath, 1996: 229–230). Other republics, freed from Cold War strategic restrictions upon their markets, had natural uranium to burn. Following suit, US surplus stocks of HEU began to be blended down, so accelerating the whole process one further degree.

While, in theory, none of this was meant to disrupt the existing commercial market, it was notable that already low uranium prices began to fall below the cost of production in all but the most efficient mines. And bargain basement prices allowed current uranium production to fall well below the forward supply requirements for LEU fuel going into established reactors – to say nothing of new ones that might, one day, be built. It eventually became evident that the process of clearing the market would have to include a considerable price rise needed to bring on new production – and indeed, prices are already notably higher than they were just a few years ago, when talk of a nuclear renaissance seemed particularly remote (Hore-Lacey, 1999).

The new Australian mines have, therefore, always been competing against stock draw-downs of one or another kind, and seem liable to do so for some time yet. The balance of the half-done HEU deal will see the LEU equivalent of about 12,000 tons of yellowcake arriving annually in western markets until 2013 – a volume that has yet to be reached by Australian exports (Steyn, 2004: 12–13). Furthermore, the end of the current HEU deal may simply witness the birth of a successor. And even if it does not, Russian authorities might simply choose to sell their blended down HEU directly into the market rather than through the agency of the US government. Either way, current and future dynamics are testament to a kind of market failure that comes from the opposite end of the spectrum to oil; instead of arising from the triumph of liberalism, it is the pervasiveness of government intervention at all stages in the history of uranium that has constituted the central flaw in this market. The local industry has skilfully avoided extensive mention of this, while getting about the job of creating a public perception of great economic injustice inflicted through the discrepancy between Australia's pre-eminent share of known low-cost ores and its much smaller proportion of current production.[16] But in this glutted market, the threat and periodic actuality of Australian capacity expansion has been one of the forces contributing to low prices – and a powerful commonsense rationale, one might add, for something like a supply-restricting federal position generally akin to the three-mines policy. Instead the three-mines policy has been depicted as holding the industry back from realising its full sales potential, while the firms in the industry have themselves sought to profit in small ways from the glutted market (Fitzgerald, 1992).

Given this obtuse perversity in the inherited structure of domestic uranium debates, it is hardly surprising that there is little public appreciation of the complexity of the uranium market. This ignorance will persist even as prices begin to rise, fed by the first significant plans for nuclear expansion in a long time.[17] In combination with greenhouse worries about the future of coal, the uranium mining industry everywhere is now talking about 'a nuclear renaissance'. Geographically and geologically speaking, the Australian industry appears to be positioned right in the front row, with aggressive plans for the capacity expansion of existing mines and the creation of new ones matching Beijing's intentions to build two reactors per annum for the next 15 years (plus Indian plans rated around half that scale). Taken together, these two national plans represent the most bullish nuclear power programme since the Japanese built more than fifty reactors in the two decades following the first oil shock.

India and China may, however, create the misleading appearance of a global trend, for the nuclear industry elsewhere appears to have sunset rather than sunrise characteristics; there are too few new reactor orders elsewhere, and age has not wearied the memories of past accidents (Schneider & Froggatt, 2004). Nonetheless, the bandwagon of interests aligned behind the latest uranium renaissance in Australia is breathtakingly broad. All possible nuclear activities short of the bomb option – mining, enrichment, high-level waste storage, even domestic power reactors – have been publicly placed upon the table by someone or other, with the government selling seats around its edge to all comers – including, most important of all, environmental opponents of the carbon age. Whatever else this might portend, it at least seems clear that the Australian uranium export industry will emerge from these debates in expanded and unrestricted form, ready to have a real tilt at the number one spot in the market enjoyed by Canadian producers.

Roxby alone is now commonly estimated to hold one third of known world low-cost uranium reserves.

But there are nonetheless delicate political issues arising from the interactions between domestic and international developments, and they are inevitably under-estimated by those wielding a new broom approach to supply-side liberalisation. One of these issues concerns the domestic refraction of Australian uranium sales to nuclear weapon states. In 2006, the Australian government agreed to export uranium to China, after Beijing had negotiated a set of safeguards with it. The last time an Australian government moved to sell nuclear materials to a nuclear weapon state – the sales to France by the Hawke government in the 1980s – it quickly produced considerable egg on the Australia's face. Earlier claims about the physical separation of France's civil and military fuel cycles were suddenly shown to be false, so blowing the bottom out of the rhetorical form in which Australian safeguards were publicly marketed, and forcing the government to regroup around 'the principle of equivalence' (Leaver, 1988).

While Canberra was making something of a fool of its own principles in public, France was at least bound by the safeguard protocols of the Euratom system. China, by contrast, is nothing more than a recent member of the NPT and its associated Nuclear Suppliers Group – and the NPT does not require its nuclear weapon states to open up their civil fuel cycles to the International Atomic Energy Agency inspec-torate.[18] Canberra's traditional argument that Australian uranium sales are adding to the integrity of NPT safeguards therefore shrinks quite considerably: ten times nothing is still nothing. Canberra, in short, is sitting down to bargain with China in circumstances of not-so-splendid isolation where the pre-existing inequalities of raw bargaining power are massive, the collective desire of industry and government for enhanced sales has been well telegraphed, and the backdrop of international and regional safeguards is non-existent. And should Canberra sit down with India, then one of the fundamentals of three decades of Australian policy – the refusal to sell to non-NPT members – will have already been mortgaged.

All this highlights the intensely political question of whether the results of a successful Australian conquest of uranium market shares can be sold back home. Thirty years ago, the international re-entry of Australian uranium was made possible through a compensatory ratcheting up of Australian non-proliferation diplomacy. The current expansion is, conversely, being contemplated at a time when the whole NPT system confronts a challenge to its survival that is without precedent. Gut feeling says that the prospect of economic rewards alone will not suffice to carry the domestic debate, and especially so when images of nuclear anarchy are likely to be increasingly easy to sustain (and even easier to link back to Howard's 'new dispensation' that combines terrorism with WMD). If so, then the question of what new diplomatic thrust along the strategic front might be required to legitimate enhanced uranium sales needs addressing. Perhaps the time has come again for reconsidering PACATOM-like regional fuel cycle initiatives? But perhaps this is altogether too tame? If we are really talking about uranium sales to India and China, then what about linking questions of supply to the Fissile Material Cut-Off Treaty?

Gas: the domestic blockage

The commercial development of North-West Shelf gas has been a long time coming, with the discovery of the major fields extending right back to Australia's pre-OPEC resources boom. And its protracted development process ties these deposits to some of the epic stories of Australian politics: the loans scandal that ultimately brought down the Whitlam government was, most notably, set in motion by that government's desire to build the overland pipeline network that would bring North-West Shelf gas to the boomerang coast. In fact domestic sales out of the North-West Shelf are today still restricted to Perth, which is very small beer compared to the booming export sales of LNG that commenced in the late 1980s to Japan (Harman, 1994). Recent deals for supply into southern China, with further prospects in California and Chile, have positioned this geological province at the highest levels of the relatively new international trade in LNG.

All this export success begs a domestic question that infrequently asked: if North-West Shelf gas is capable of penetrating distant electricity generating markets, then what stops it from entering domestic ones? Part of the historic answer is that the original visionaries thought of this gas primarily as feedstock for a petrochemical industry. But the begged question achieves renewed salience in the context of greenhouse warming, for the level of carbon emissions from gas burning is less than half that of coal – not a silver bullet by any means, but a quick, cheap and readily available interim reduction nonetheless. Another part of the answer to the question used to be sheer physical isolation, but this is beginning to wear thin for two reasons. First, something like the national system of gas pipelines is today within relatively cheap reach, largely as a result of the ad hoc linkage of smaller gas fields to major cities. This state-centric 'growth pole' pattern of development has allowed gas to become a highly successful domestic fuel source, and its share of national primary energy is already growing much faster than any other energy form. And perhaps most importantly, all state capitals save for Perth are now networked, albeit with Adelaide and Brisbane connected into the eastern seaboard rather indirectly through the Moomba hub (ABARE, 2004: 24). Second, the cost of transporting LNG by ship has lowered dramatically over the last decade, placing LNG in direct competition with long-distance pipelines (Lee, 2005: 1). If domestic gas and North-West Shelf LNG already have independent bright futures, then it is not difficult to imagine how much brighter they would both be if they were linked to each other in one of the above-mentioned ways.

But the slow pace of forward movement along the front of national integration has always spoken to more than just physical isolation. It also highlights the close integration between localised coal mining and electricity generation that was traditionally sustained by state-level governments. With the highest grades of thermal coal committed to export, Australian electricity generation fell back upon lower grades that had dubious thermal and environmental qualities. State monopoly powers were particularly important during the second age of energy security, when the booming electricity demand associated with suburbanization and energy intensive industries frequently allowed the former end-use to covertly subsidize the latter. The environment, of course, lost on both counts.

A clear window upon this heady brew of issues is provided by the July 2005 decision of the Gallop Labor government in Western Australia to award a new base-load generating contract to a gas-fired tenderer (Drummond, 2005). Historically, the state government electricity monopoly sourced energy for its base-load generators from state-owned coal mines where the grade of coal was moderately low. Even allowing for pipeline costs, North-West Shelf gas had little trouble undercutting the coal price in the Perth market, and the export-oriented bauxite/alumina industry south of Perth was very quick to take advantage of this differential. What gas could not do, however, was politically out-muscle the established players in the base-load generating game, especially when Labour governments (with their close connections to coal unions) were in office. So despite an explicit recommendation from a 1989 Power Options Review Committee to go with gas as the cheapest and cleanest fuel, coal continued to dominate the state electricity scene. The Gallop government decision to finally switch over to gas therefore has some of the qualities of a historical epic to it. In sheer cost terms, it ought nonetheless to have been an open-and-shut case, for the combination of lower fuel and capital costs associated with gas now confers a whole order of magnitude advantage. Nonetheless, the strength of the political push to reverse the July decision dramatically shows how bitter the infighting can be – as it would be if North-West Shelf gas were ever fed into the publicly owned electricity generating systems of most eastern states.

Coal: once again, the fuel of the future?

It was thermal coal more so than any other segment of the Australian energy scene that had greatness thrust upon it during energy security's second age. Australia soared up the ranks of coal exporters like the proverbial rocket, emerging as the world's largest coal exporter in just over a decade (Ekawan, Duchêne and Goetz, 2006). The boost phase was, of course, provided by the Japanese retreat from over-dependence upon expensive oil, with the Tiger economies following close behind. But it is too easily forgotten that none of these economies had an industrially significant domestic coal industry when Australian supplies marched into their energy markets. The challenge for coal exports in the third age of energy security is to replicate something like this trajectory with regional customers that already have high coal dependence, and regard it as part of their energy and environmental problems. Under these conditions, Australian re-runs of its Japanese story will fail to entertain in Beijing and New Delhi, with or without the moral tutelage of the Kyoto Protocol. Recent price rises for thermal coal merely reconfirm that teaching.

The status of being the world's largest exporter also confers an unusual challenge upon Canberra and the Australian coal sector. When energy and coal exports were relatively small, then Australian attitudes could be those of the opportunist free rider hitch-hiking upon markets whose dynamics were determined in other places. But this 'small country assumption' progressively became less appropriate as exports expanded, coming into conflict with the emergent large country realities. And failure in the marketplace for self-images then prefaced the profound sense of denouement – evidenced earlier in coal's profitless growth syndrome and Doug Anthony's uranium heroics. Oddly enough, this challenge is one that the Howard government has rhetorically nodded at inside the strategic sphere. Even before day one, prospective Foreign Minister Downer insisted that Australia was 'more than a middle power' – a line of argument that represented a radical abandonment of the small country assumption behind Menzies' foreign policy (Downer, 1996, 2003). But what has been absent to date, however, is any analogue to that theme in relation to energy markets in general and coal in particular.

This context of missing ideas provides what is perhaps the kindest way to think about the Asia–Pacific Partnership for Clean Development and Climate (APPCDC) launched on the back slope of the July 2005 ASEAN summit in Vientiane. From the beginning, there was a variety of less kind ways, and a good portion of the national and international press was full of them. In truth it was much too soon to be certain about ultimate purposes, for the partnership consisted of little more than a Vision Statement, a set of important signatures and an agreement to meet. The Vision Statement, furthermore, nodded towards the Whole Earth Catalogue of abatement measures and technologies from efficiency drives to carbon sequestration to nuclear power and renewables (Joint Press Release, 2005). The good news in it was that the signatories at least felt the need to recognise the reality of climate change,[19] and that this collection of big coal producers and consumers seemed to be committed to promoting the R&D investments that will be necessary to get greenhouse gas reductions down by the 60 per cent required this century. But the central problem that Canberra wanted the APPCDC to address was the same one that the Howard government highlighted in its own 2004 Energy White Paper – namely, worries that the quantum of investment in new coal-based generating facilities would not be forthcoming in a timely fashion, and that failure along the investment front would fold back into lower horizons for coal exports (Australian Government, 2004: 136 and 144, respectively). Coal-fired power stations have long life-spans and are relatively expensive to build, so that clear views of coal's future out to the 50-year time horizon of modern plants is already intersecting alarmingly with negative environmental feedbacks. So geo-sequestration always appeared very much the central Australian interest in this pact.

The danger, of course, was that the APPCDC partners, though modest in number, had signed on to a process that would not ultimately yield substantive content. The Kyoto Protocol, with which APPCDC is said to be complementary, also has a clean development mechanism, and both India and China have hitherto expressed considerable policy interest in it as a way to obtain foreign investment in renewables. Unlike the Kyoto Protocol, however, the APPCDC Vision Statement mentions no timetables, targets, mandatory caps or penalties. Furthermore, the intended initial meeting of its signatories clashed rather pointedly with the first-ever meeting between the G-8 and key-developing countries which had been arranged under Blair's leadership at the Gleneagles Summit. Given the mixed interests of India, China and Japan,[20] there was a realistic prospect that the first APPCDC meetings would not agree on anything too grand – but just grand enough to establish the promise of forward movement on geo-sequestration. Over the short run, when promise is often more powerful than achievement, it may be sufficient to bring forth the utility investments that will lock in forward commitments for future streams of coal purchases. If the results of future APPCDC meetings fall into the same 'potential gap' that enveloped the inaugural January 2006 meeting, then it will be tempting to label the organization as 'the Shane Stone approach' to climate change: mean and tricky.

Conclusion

It is not difficult to detect a presumption leaning in the Shane Stone direction from the manner in which the above comments have been framed. This presumption, it is worth saying, has very little to do with an appreciation of the alleged costs and benefits of the Kyoto Protocol as opposed to any other mechanism for achieving climate change, and everything to do with watching the 'gain without pain' philosophy that the Howard government has applied to climate change policy over the last decade. Having initially fought hard for a special and highly favourable increase in Australia's target for carbon emissions under the Kyoto Protocol, the government's subsequent unwillingness to seek parliamentary ratification for its own signature showcased a degree of deviousness not evidenced in Australian diplomacy since the Gorton government danced around the NPT in the early 1970s.[21] This deviousness has also highlighted the extent to which the multinational-dominated export coal industry has been the *eminence grise* to this diplomatic double play, a role that rests upon its unstated threat to redirect future investments towards countries outside the protocol. All of this is, in turn, highly consistent with other important parts of the domestic pattern of energy policy, particularly the low and shrinking federal commitment to renewable energies. So while it takes considerable hard work to negate the presumption of good faith that normally accompanies diplomatic initiatives, on this score the Howard government has not flagged.

If there is any single linchpin that attaches the fate of the APPCDC to the market failures that lie latent within Australia's third age of energy security, it is the price of carbon. Given the diversity of Australian primary energies, it would be difficult even under the most favourable of circumstances to locate the proverbial goldilocks position where the price of carbon was just right, since different energy sub-sectors tend to pull that price towards opposing extremes. With zero emissions during operation, nuclear energy in particular profits from high carbon prices, since this would more than offset the substantial up-front capital cost of nuclear projects. Although its carbon emissions are considerably higher, gas would possibly benefit even more than nuclear in the Australian context, since it already enjoys a quite handsome profitability over low-grade domestic coals that would be further enhanced by a high carbon price.[22] Coal and liquid fuels, on the other hand, have much brighter horizons insofar as the price of carbon is regarded as an externality, and in this respect these industries can generally draw short-term political support from the workers and consumers whose fates are tied to those technologies. Keeping nuclear power exiled from the domestic energy scene, and confining Australia to a nuclear future as a raw materials trader, is therefore an important perimeter defence to the domestic and international positions enjoyed by Australian coal.[23]

In this context, it is notable that the most recent IEA review of Australian energy policies strongly recommended that the Howard government develop a price for carbon. Equally notably, Howard's environment minister was quick to say that a carbon price was a conceivable product of the APPCDC process – but not for a considerable time (Peters, 2005). So as Australians head off toward a new debate about the future of different energies, there is much that hangs in the balance, both at home and abroad.

Notes

1 The extent to which there was the perception of a radical break has dulled over time. Nonetheless, the best of this literature remains the work of Geoffrey Barraclough, the master craftsman of conjunctures and disjunctures; see in particular Barraclough (1974).

2 This instance of nuclear co-operation moved into higher gears during the Suez War of 1956 – initially as a side-agreement to the Protocol of Sévres that put the war on track, and subsequently in reaction to the overt Soviet nuclear threat directed at Israel during the war. For pertinent reflections, see Pinkus (2002).

3 The reprocessing deal was eventually abandoned under US pressure in the early 1980s, although it was commonly thought that 95 per cent of the technology had been transferred by that time; see Dorian and Spector (1981: 59–60).

4 For one of the more comprehensive investigations into the growth of suspicions, see Snyder (1985).

5 For one analysis of the political repercussions, see Imber (1989: Chapter 5).

6 For the policy expressions of this enhanced scepticism, see Feldman (1982: 183–184).

7 China, of course, also took up its NPT seat at more or less the same time, so ending what was truly the most dangerous phase in the history of nuclear commerce.

8 Even its most advanced phase – the 1974 flirtation of the Dunstan Labour government in South Australia with the idea of nuclear enrichment – was really just an elaborate example of value adding through minerals processing.

9 Coal was not, of course, the only energy pathway open to Japan's utilities. They also invested heavily in nuclear energy. A decade later, both would make roughly equal contributions to Japan's primary energy supply.

10 For analysis of the early stages of this transformation, see Fagan (1981), and especially p. 154, where it is reported that electricity prices in the major Australian states were less than a quarter of utility rates in Japan.

11 This popular rejection was, paradoxically, fuelled by the Coalition governments of the 1960s, all of which did their best to publicly discredit the integrity of the NPT system of safeguards that was then being negotiated. At the time, they hoped to create a solid social foundation for Australia's later rejection of the NPT and a possible take-up of the nuclear weapons option.

12 These debates raged over the whole decade from 1975, when they constituted one of the two largest issues on the national political agenda: see Smith (1979). Even so, Australia's uranium debate was almost entirely concerned with the manner in which uranium exports interfaced with non-proliferation rather than energy security objectives. And very little was heard in all this about the interface with Carter's policy – in part, perhaps, because Carter's policy did not survive him. Reagan was forced to accommodate to a political victory by his allies over the advanced fuel cycle (although, ironically, the fullness of time would reveal that this political victory did not necessarily translate into a technological victory. The sorry fate of the breeder reactor in France and Japan is a case in point).

13 Calder was writing, of course, before the Asian Financial Crisis, but this crisis ultimately left at least the resource dependency parts of his thesis untouched.

14 Consider, for instance, the role of the SARS crisis in pushing China up to the number two slot on the league table of national oil consumers. At the time that crisis erupted, SARS was thought likely to reduce the level of economic activity in China, including in the car industry, which was regarded by the Party as an absolutely central sector in the nation's industrial transformation. But in fact it had precisely the opposite short-term effect for the auto sector. Consumers who could afford to do so shunned public transport and brought forward their purchases of cars. Hence the annual rate of increase of auto sales, previously a very heady 30 per cent, jumped to the absolutely phenomenal pace of 80 per cent in the 12 months following the SARS outbreak. Since this coincided with the completion of a truly national system of multi-lane freeways, that venerable marketing slogan from the age of the infant Japanese auto industry – 'freedom is a Honda' – seems well placed to experience a second Chinese coming.

15 'Practical' and 'realistic' have long been Downer's highest commendations. But upon becoming Australia's longest serving foreign minister, he marked the occasion by extending his argument to the point where 'vision' became anathema. So 'Stalin had a vision, Hitler had a vision' – while for Downer, mimicking George W. Bush, 'freedom' was everything: see Starick (2004).

16 These two ratios are currently quoted at around 40 and 20 per cent, respectively, whereas for most of the period up to the mid-1990s, they stood around 30 and 10 per cent, respectively. Roxby expansions account for most of the shift.

17 During its first 12 months in office, the Howard government appeared to be on the verge of clearing away Labour's three-mines policy. It was preparing itself for Indonesia's aggressive reactor-building programme sponsored by Suharto's Minister for Technology, B.J. Habibie. But by the time those preparations emerged from Democrat-inspired probings in the Senate, Indonesia and its grandiose nuclear plans had unfortunately disappeared over the abyss of the Asian financial crisis. The government therefore decided to let the sleeping dog of a public uranium debate lie.

18 Nonetheless, at different points in time, all of the nuclear weapon states have decided to open up some of their nuclear facilities to IAEA inspection. But these 'voluntary offer safeguards' are, as their name suggests, voluntary rather than mandatory.

19 Not all, however. For an eloquent articulation of the old orthodoxy by the first-ever US energy secretary, see Schlesinger (2005).

20 At Davos in February, China announced (in the presence of John Howard) that all their new power stations would henceforth use either hydro or nuclear technologies; see Gottliebsen (2005).

21 In that instance, Australia's signature was accompanied by statements that ratification would not proceed until 'matters of concern' laid out over the previous 5 years had been rectified, and that future Australian withdrawal from the treaty was possible if these matters were not attended to. What was not at all apparent until 5 years later was that the nuclear reactor that the government was simultaneously building was never intended to be covered by NPT safeguards. Australia's signature was, therefore, little

more than a cheap moral scutcheon designed to take advantage of the unusual legal feature whereby those who signed the NPT before it entered into force were not bound by their signature, while those who signed after entry into force were. And Australia provided the thirty-ninth of the forty signatures needed to activate the treaty. On this episode, see Walsh (1977: 13).

22 In 2002, the OECD's Nuclear Energy Agency (2002: 36) reported a 'rule of thumb' whereby the increase in costs experienced by gas-fired and coal-fired power plants subject to carbon taxes would be in the ratio of 2 to 5. The ratio would, of course, tend to blow out with lower grades of coal – like the grades that dominate domestic consumption in Australia.

23 For one of this theme's more lively articulations, see the remarks of Queensland premier Peter Beattie, as cited in Viellaris (2005).

7 Energy security: Pacific Asia and the Middle East

Gawdat Bahgat

In January 2005 an unprecedented meeting between major Persian Gulf oil producers and Asian oil consumers was held in the Indian capital New Delhi. Energy ministers from Iran, Kuwait, Oman, Qatar, Saudi Arabia, the United Arab Emirates, China, India, Japan, and South Korea discussed different proposals to consolidate oil and gas cooperation between the two sides. These included the development of an Asian petroleum market, mutual investments in upstream and downstream sectors, and building strategic petroleum storages. This gathering underscores a fundamental characteristic of the global oil market – the Persian Gulf states are the major centre of gravity on the supply side while Pacific Asian nations have become the major centre of gravity on the demand side of the equation.

The implications of this growing partnership on global energy markets and strategic ramifications are the main focus of this chapter. The framework of analysis can be summarized as follows. For the next decade global oil markets will continue to reflect competition between three major producing regions – former Soviet Union (including Russia and the Caspian Sea), West Africa, and the Middle East. On the consuming side the competition is mainly between the Pacific Asia, the United States, and Western Europe. Each of the consuming regions has forged an energy partnership with one or more of the producing areas. Europe receives substantial proportion of its oil and gas supplies from Russia. Since the early 2000s the United States has sought to reduce its dependence on oil supplies from the Middle East and sought to increase imports from Canada, Mexico, Venezuela, and West Africa. Finally, Pacific Asia's skyrocketing demand is met, mainly, by supplies from the Persian Gulf. In the foreseeable future the Middle East, particularly the Persian Gulf producers, will continue to be the driving force to ensure global energy security. The world will grow more dependent on oil supplies from the Middle East. The region has the hydrocarbon resources to meet growing global demand.

The World oil market is well-integrated. Competition between various producers and consumers should not be seen in zero-sum terms. The source of oil matters less than the availability of supplies. In other words, in today's oil market who buys and who sells one barrel of oil has little impact on energy security. Instead, the availability of adequate supplies significantly insures security and stability. Within this context, the developing energy partnership between Asian Pacific nations and Gulf producers should be seen as a positive step. It would enhance energy security for both sides and would contribute to global economic stability and prosperity.

The following section examines the concept and implications of 'energy security'. This will be followed by a brief analysis of Russia's, the Caspian Sea's, and West Africa's oil potential. The following section will discuss the main characteristics of the energy sector in the Gulf producers and Pacific–Asian consumers. Finally, the strategic environment and the geopolitical ramifications of the Asian/Gulf energy partnership will be examined.

Energy security

Modern society has grown more dependent on energy for almost all human activities. Different forms of energy are essential in residential, industrial, and transportation sectors. Energy is also crucial in carrying out military operations. Indeed, the attempt to control oil resources was a major reason for the Second World War. In short, our increasing reliance on energy has heightened the importance of energy security. The first oil shock in the aftermath of the 1973 Arab–Israeli war put energy security, and more specifically security of supply, at the heart of the energy policy agenda of most industrialized nations (LaCasse & Plourde, 1995: 1). Since then policy-makers and analysts have sought to define the concept 'energy security' and its implications.

The European Commission defines energy security as 'the ability to ensure that future essential energy needs can be met, both by means of adequate domestic resources worked under economically acceptable conditions or maintained as strategic reserves, and by calling upon accessible and stable external sources supplemented where appropriate by strategic stocks' (cited in Skinner & Arnott, 2005a). Barton *et al.* (2004: 5) define it as 'a condition in which a nation and all, or most, of its citizens and businesses have access to sufficient energy resources at reasonable prices for the foreseeable future free from serious risk of major disruption of service'. In short, energy security refers to sustainable and reliable supplies at reasonable prices. In this essay the concept of energy security includes the following parameters. First, the different threats to energy security include geopolitical, economic, technical, psychological, and environmental ones. Second, the definition of 'security' embodies the element of 'price' or achieving a state where the risk of rapid and intense fluctuation of prices is reduced or eliminated. Oil prices vary from country to country depending on several factors including the quality of crude, destination, taxes, exchange rates, and refining capacity, among others. For a long time the Organization of Petroleum Exporting Countries (OPEC) has played the role of swing producer. This means that when other producers, such as Russia, the Caspian, or West Africa, increase their production, OPEC reduces its share in order to prevent prices from falling. In addition, since the early 2000s OPEC has adopted a 'price band', which reflects the organization's preferred price range.

Third, prices have a strong impact on the availability of funds to invest in exploration and development of oil resources. Energy security depends on sufficient levels of investment in resource development, generation capacity, and infrastructure to meet demand as it grows. Traditionally, high oil prices have led to accumulation of funds in the hands of national and international oil companies and more investments. Eventually, new investments add more supplies to the market and contribute to lower prices. Systematic under-investment characterized the oil industry in the 1990s due to stable oil prices at low levels since the mid-1980s. This under-investment contributed to a shortage of supplies and higher prices since the early 2000s. Fourth, spare capacity has traditionally played a significant role in temporary severe interruptions of oil supplies. A few OPEC producers, particularly Saudi Arabia, have purposefully maintained spare capacity to ensure stability in global markets. Global economic growth, particularly in Pacific Asia, has subjected the oil market to an unexpected demand shock that has practically eliminated spare capacity. Accordingly, the international oil industry has entered a period of fundamental change. In the mid-2000s spare capacity is at one of its lowest recorded levels.

Fifth, security of supplies can be enhanced by an overall diversification of supply. Put differently, the more producing regions the more stability in international oil markets. Thus, increasing supplies from Russia, the Caspian Sea, West Africa, and other regions would reduce the vulnerability of over-dependence on one single region. Wars, military operations, and political tension in the Middle East have prompted calls to reduce dependence on supplies from that region. Although the political situation in the Middle East provides many grounds for concern there has not been any major disruption of supplies from the region since the 1973–1974 oil shock. Middle Eastern producers realized that imposing oil embargo for political purposes was unproductive. In the following decades major producers increased their production to compensate for any shortage resulting from political upheavals around the world. Finally, from the perspective of producers, demand security also merits attention. Major resource holders have voiced their concern regarding long-term security of demand for their oil (Lajous, 2004). This concern is based on two grounds: a) the cyclical growth patterns and policies that dampen the demand for oil and favour other sources of energy; and b) OPEC producers have failed to diversify their economies and continued to be heavily dependent on oil revenues. Thus they are concerned about securing markets for their major source of income. Within this context, the growth of Asia's oil demand is seen as a welcome development by OPEC producers.

To sum up, the globalization of the oil market suggests that rhetoric regarding the goal of self-sufficiency in energy is obsolete. Energy security is an international issue that requires growing interdependence between major producers and consumers. The skyrocketing demand for oil in Pacific Asia is a case in point.

Pacific Asia

A recent report issued by the United States National Intelligence Council (NIC) predicts that the twenty-first century is likely to be the 'Asian Century'. It argues that the emergence of China, India, and other Asian powers, is similar to the advent of a united Germany in the nineteenth century and a powerful United States in the early twentieth century (NIC, 2004: 9). This rapid rise of Asia is driven by the incredibly high and sustained economic growth the region has witnessed since the early 1990s, led by its fastest growing economies – China and India. Together the two nations have more than 2.2 billion people, more than one-third of total world population. Finally, both China and India are nuclear powers and their military capabilities, both conventional and non-conventional, are on the rise. In short, Asia has the necessary 'ingredients' to become a global power and already is on its way to becoming a prominent player on the international scene.

Several characteristics can be identified in Asia's energy outlook. First, Asia–Pacific nations have very limited oil and natural gas reserves. Indeed, as Table 7.1 illustrates, the region holds the lowest oil reserves and the third-lowest natural gas reserves in the world.

Second, Pacific Asia has a huge and growing gap between its low oil and gas production and its consumption as Table 7.2 shows. The figures show variation between countries but Pacific Asia as a region produces only 10.2 per cent of world oil and consumes 28.8 per cent. The figures for natural gas are 11.9 per cent and 13.3 per cent, respectively. These gaps are filled by imports from overseas, particularly from the Persian Gulf. Third, like in Europe and the United States, a substantial proportion of oil in Pacific Asia is used to meet the region's rising transportation sector, particularly in China and India. Their sustained high economic growth has led to soaring in vehicle ownership. The number of vehicles in China in 1980 was less than two million. By 2002 it increased ten-fold to almost 18 million. In India, vehicles totalled 10.7 million in 2000, an increase of 245 per cent since 1984 (Cannon, 2005). In addition to these general characteristics, a brief examination of the region's four largest economies (China, India, Japan, and South Korea) will shed light on energy resources and policies in Pacific Asia and the implications on global energy markets.

Probably more than any country in the world, China's skyrocketing demand for energy, particularly oil, has significantly shaped the global oil market since the mid-1990s. China is both the largest consumer and producer of coal in the

Table 7.1 Oil and natural gas reserves 2004

Region	Thousand million tonnes	Thousand million barrels	Share of world total (%)	Trillion cubic feet	Trillion cubic meters	Share of world
Asia–Pacific	5.5	41.1	3.5	501.5	14.21	7.9
North America	8.0	61.0	5.1	258.3	7.32	4.1
Europe	19.0	139.2	11.7	2259.7	64.02	35.7
South America	14.4	101.2	8.5	250.6	7.10	4.0
Africa	14.9	112.2	9.4	496.4	14.06	7.8
Middle East	100.0	733.9	61.7	2570.8	72.83	40.6

Source: British Petroleum. (2005). *BP Statistical Review of World Energy*, 4, 20.

Table 7.2 Oil and natural gas production and consumption in Pacific Asia, 2004

Country	Oil production	World share (%)	Oil consumption	World share (%)	Gas production	World share (%)	Gas consumption	World share (%)
Australia	541	0.6	858	1.0	35.2	1.3	24.5	0.9
Bangladesh	NS	–	86	0.1	13.2	0.5	13.2	0.5
Brunei	211	0.3	NS	–	12.1	0.4	NS	–
China	3,490	4.5	6,998	8.6	40.8	1.5	41.2	1.6
India	819	1.0	2,555	3.2	29.4	1.1	32.1	1.2
Indonesia	1,126	1.4	1,150	1.5	73.3	2.7	33.7	1.3
Japan	NS	–	5,288	6.4	NS	–	72.2	2.7
Malaysia	912	1.0	504	0.6	53.9	2.0	33.2	1.2
New Zealand	NS	–	151	0.2	3.6	0.1	3.6	0.1
Pakistan	NS	–	296	0.4	23.2	0.9	25.7	1.0
Singapore	NS	–	748	1.0	NS	–	7.8	0.3
S. Korea	NS	–	2,280	2.8	NS	–	31.6	1.2
Taiwan	NS	–	877	1.1	NS	–	10.1	0.4
Thailand	218	0.2	909	1.2	20.3	0.8	28.7	1.1
Vietnam	427	0.5	NS	–	4.2	0.2	NS	–
Others	184	0.2	747	0.9	18.2	0.7	11.3	0.4
Total Pacific Asia	7,928	9.8	23,446	28.9	323.2	12.0	367.7	13.7

Figures for oil are in thousand barrels and for natural gas in billion cubic meters.

Source: British Petroleum. (2005). BP Statistical Review of World Energy, 6–25.

world and coal makes up the bulk of its primary energy consumption. However, Beijing is increasingly switching to oil and to a lesser extent natural gas. Unlike most industrialized countries, China did not have to deal with the economic and political consequences of the 1973 oil shock. The nation was self-sufficient. This luxury did not last long. In 1993 China became a net oil importer and a decade later, 2003, it surpassed Japan to become the world's second largest petroleum consumer. The reason is domestic production could not keep pace with rising demand.

In order to fill this widening gap between declining domestic production and rising production Beijing has sought to re-structure its oil industry and to sign oil deals with as many producers as possible. The re-structuring of the country's oil industry was further accelerated with China's entry into the World Trade Organization (WTO) in late 2001. The Chinese government committed itself to broad economic liberalization. In the mid-2000s four companies dominate China's oil industry. These are China National Petroleum Corporation (CNPC), China Petrochemical Corporation (Sinopec), China National Offshore Oil Corporation (CNOOC), and Sinochem.[1] These national companies face growing competition with international oil companies. With the market liberalization required by the WTO, China amended regulations governing foreign investment in both its onshore and offshore oil and gas sectors. Most foreign investment has been concentrated in offshore exploration and development where Chinese companies have limited experience. Finally, until the 1990s natural gas was used largely as a feedstock for fertilizer plants, with little use for electricity generation. The drive for diversification of the energy mix and concern about pollution have prompted the Chinese government to embark on a major expansion of its gas infrastructure and to secure natural gas supplies from overseas.

In order to secure oil and gas supplies from abroad Chinese officials and oil executives have negotiated deals with producers all over the world including Canada, Venezuela, Iraq, Iran, Saudi Arabia, Sudan and Yemen. Chinese companies have been particularly interested in countries such as Iran and Sudan where they do not have to compete with their American counterparts.

India's economic and energy outlooks are similar to those of China. In recent years the Indian economy has experienced high growth rates, albeit lower than China. This economic growth was the result of a broad and extensive economic reform program. India, like its giant neighbour, depends on coal as the main source of energy and is increasingly switching to oil and natural gas. With much more limited hydrocarbon resources, India, more than China, is heavily dependent on foreign supplies to meet its growing demand for oil and gas. Again, domestic supply has lagged behind growing demand and the gas is filled by imported oil and gas mainly from the Persian Gulf.

Unlike its two large Asian neighbours, Japan experienced slow economic growth for most of the 1990s and the early 2000s. This economic stagnation led to restrained energy demand. In the mid-2000s, the Japanese economy seems to have regained momentum and is on the way to healthy economic growth. This recovery is likely to increase the demand for energy. This is particularly alarming given the country's lack of significant domestic sources of energy. Despite its large economy and high standard of living, Japan is one of the poorest industrialized countries in energy resources and, consequently, is highly dependent on foreign supplies. For several decades Japanese oil companies have been actively involved in oil and natural gas exploration and development projects all over the world. Given the country's small coal reserves and Japan's international stand against pollution, natural gas plays a significant role in its energy mix. Indeed, Japan has been a pioneer in importing liquefied natural gas (LNG).

South Korea is similar to Japan in terms of lack of indigenous hydrocarbon resources. Seoul depends almost completely on foreign supplies to meet its oil and gas needs. Along with the United States, China, Japan, and Germany, South Korea is a leading oil importer. The state-owned Korea National Oil Corporation (KNOC) is pursuing equity stakes in oil and gas exploration around the world. It has several exploration and production projects in Yemen, Libya, Argentina, Peru, Venezuela, Vietnam, and Kazakhstan (EIA, 2005c). South Korea also relies on imported LNG for most of its natural gas needs.

To sum up, despite some variation between the four largest Asian economies they all share a dominant characteristic – domestic oil and gas reserves are extremely limited and inadequate to meet their current and anticipated economic growth. They compete with each other and with other major consumers over hydrocarbon resources. The three most prominent targets are West Africa, Russia and the Caspian Sea and the Persian Gulf.

West Africa

West Africa, especially its offshore areas, which are believed to be very promising petroleum regions, has become a major oil-producing province, where most of the international oil companies are actively working in development and production projects. African petroleum is particularly prized because of its high quality – light and sweet (low sulphur). Most of the African oil comes from the two top producers – Nigeria and Angola.

Nigeria is the world's seventh largest oil producer. A significant challenge to the full utilization of the country's hydrocarbon resources is political instability. Since its independence from the United Kingdom in 1960 Nigeria did not hold successful elections under a civilian government until the late 1990s.[2] The election of President Olusegun Obasanjo in 1999 (he was re-elected for another term in 2003) provided a hope for stability. The federal government, however, has been in conflict with regional state governments over control of the country's offshore oil and gas resources. The former wants to maintain its ownership and control of all natural resources within territorial waters, while the states want more of the oil revenues to be allocated to them and not to the federal government. Despite intense negotiations and signing several agreements a compromise has yet to be reached. Thus, political and ethnic strife in the Niger Delta, where the majority of oil reserves are located, often disrupts Nigerian oil production. This includes kidnapping, seizure of oil facilities and illegal fuel siphoning. In addition to disrupting domestic production, these political upheavals contribute to volatility of global oil markets and prices.

Economic and political conditions in Angola are not any better than those prevailing in Nigeria. Angola is beginning its recovery from a devastating 28-year civil war that began shortly before the nation achieved independence from Portugal in 1975. The civil war had devastating impact on the economic infrastructure and displaced millions of people. An agreement to end the civil war was finally reached in April 2002. Angola is sub-Saharan Africa's second largest oil producer behind Nigeria, with the majority of its crude oil production located offshore and in its northern Cabinda province. This province faces a situation similar to the Niger Delta states in Nigeria. Cabinda produces more than half of Angola's oil and accounts for nearly all of its foreign exchange earnings. Political tensions are high in some areas of Cabinda as separatist groups demand a greater share of oil revenues for the province's population. The separatist groups often resort to violence including sabotage and kidnapping. Meanwhile, the government has categorically ruled out the prospect of complete independence for the oil-rich, but poverty-stricken province.

Finally, Sudan which is located on the eastern part of Africa has significant oil reserves. The full utilization of these resources has been slowed down as a result of the civil war and international isolation. Chinese and Indian companies, however, have a policy of seeking oil deals in countries where they do not need to compete with American companies. In line with this policy, Chinese and Indian companies have been aggressively involved in oil exploration, development, refineries, and pipelines in Sudan.

Russia

Russia is a major player in the global oil market. Since the late 1990s, Russia's oil production has experienced a steady resurgence. By the early 2000s, Moscow had regained its status as a major oil producer and exporter and a crucial player in global energy markets. Prior to the break-up of the Soviet Union, oil production peaked at 12.6 million barrels per day (b/d) in 1987. Such high production levels stemmed largely from the exploitation of large new petroleum reserves discovered in Western Siberia. The political turmoil that accompanied the collapse of the Soviet Union was a major factor in the decline of production in the following decade. As the political situation normalized the oil industry stabilized and, gradually, production started to grow substantially. In addition to the increasing stability of the Russian political system, the introduction of economic reform and the privatization of the oil sector have contributed to this dramatic turnaround.

Russia's influence over the world oil market has risen dramatically in proportion to its growing production. The European Union has negotiated energy agreements with Russia. Moscow is a major oil and gas supplier to several European countries. The European-Russian energy dialogue is focused on European investment in Russia's oil and gas sectors, in return for steady and secure supplies. Similarly, the United States has shown growing interest in establishing an energy partnership with Russia. The goal is to reduce US dependence on oil supplies from the Middle East.

Russia's energy strategy forecasts that Pacific Asia could absorb as much as 30 per cent of the country's oil exports by 2020, compared with just three per cent in 2003. Similarly, natural gas exports to the region could surge from zero to 15 per cent of total Russian gas exports over the same period (Gorst, 2003: 10). In line with these goals top Indian officials have sought to invest in Russia's oil companies and to invite Russian investors to develop India's oil and gas industry. Chinese officials also have sought to expand their energy cooperation with Russia. The last Soviet President, Mikhail Gorbachev, put an end to Sino–Soviet hostilities in 1989, shortly before the collapse of the Soviet Union in December 1991. Despite unresolved border issues, the newly born Russian Republic and China established diplomatic relations and sought to improve co-operation in different areas. They are united by what they perceive as a unipolar American domination of the international order.

Within this strategic context, the newly established Sino–Russian friendship is built on co-operation in energy issues. Despite obvious mutual interests the two nations could not agree on building a pipeline to export Russian oil to China. Currently, most of Russia's oil comes from Western Siberia, but Moscow has ambitious plans to explore and develop oil deposits in Eastern Siberia and the Far East. This region can become a major oil provider to the Asia–Pacific market, particularly China, Japan, and South Korea. For several years Russian officials have debated two options. The two proposed pipelines will start at the Russian city of Angarsk and reach either the Chinese city of Daqing, or the port of Nakhodka. Another proposed pipeline project would link the Russian natural gas grid in Siberia to China and possibly South Korea. Several factors will influence any decision on any of these pipeline projects, most notably the availability of sufficient oil and gas deposits to make such huge investment cost-effective. Environmental considerations are also likely to be taken into account in choosing one route or the other.

Despite Russia's growing role and influence in the global oil markets, several characteristics of its oil industry need to be underscored. First, the country has a limited pool of proven crude reserves (about 6 per cent of the world's total). Major Middle East producers hold much larger reserves (about 63 per cent of the world's total). Russia's relatively limited reserves are particularly alarming, considering that Russia's rate of oil production is exceeding the rate at which new reserves are being discovered by a significant margin. Put differently, the depletion of existing oil fields in West Siberia has raised fears that Russia's current oil boom will be followed by a sharp decline in the next few years. Furthermore, most of Russia's unutilized oil reserves are located in geographically remote and geopolitically challenging fields.

Second, production costs are much higher in Russia than in the OPEC producers. The cost of production in Saudi Arabia, for example, is less than $1.5 per barrel, compared with the global average of about $5 per barrel. In Russia, it varies from one region to another, but overall it is much higher than in the Middle East.

This means that Russian firms cannot survive a prolonged period of weak oil prices. Third, given the structure of Russia's oil industry, the country does not have any spare capacity. In other words, in the early 2000s, Russia's oil industry is dominated by private oil companies. Like any private entities, these Russian companies seek to maximize their profits by producing and exporting as much as they can, with little concern about strategic objectives. On the other hand, in OPEC producers the oil industry is dominated by the state. This means that production and export policies are driven by both commercial and strategic interests.

Fourth, foreign investment has been an important component of the economic reform program which started in the early 1990s. Russian efforts to attract foreign investment, however, have been hesitant and ambiguous. As a result, the Russian economy in general and the oil sector in particular, has received a very modest amount of direct foreign investment. Available investment, domestic and foreign, will be affected by the extent to which the Russian government exerts control over private oil companies. Since the early 2000s the Russian government has sought to expand its control over the energy sector. Some analysts describe these economic measures as 'de-privatization' and 'renationalization' of oil industry.

Fifth, Russia has an extensive domestic oil pipeline system, with links to nearly all the former Soviet republics, but the country's ability to export its oil to markets beyond the borders of the former Soviet Union (FSU) is limited. This reflects the close economic ties Russia had with fellow socialist republics during the Soviet era. The break-up of the Soviet Union meant that Russia needed to expand its oil exports to Western markets in order to earn badly needed hard currency. The expansion of Russia's pipeline capacity has not kept pace with the country's rising production. Indeed, the biggest factor preventing the rapid development of Russian energy exports is its transportation network, which is exclusively under the state-owned monopoly, Transneft.

The Caspian Sea

The 700-mile long Caspian Sea is located in north-west Asia. Five countries – Azerbaijan, Iran, Kazakhstan, Russia and Turkmenistan share the Caspian Basin. Their policies on the exploration and development of the region's hydrocarbon resources since the collapse of the FSU in late 1991 have been of great interest to energy officials from all over the world. The region is important to oil- and gas-consuming countries because it can contribute to the global production and to the diversification of supplies and, consequently, reduce dependence on OPEC producers. In short, the Caspian Sea has the potential to enhance global energy security.

The region is not new to the petroleum and natural gas industry. It is worth remembering that commercial energy output began in the Caspian basin in the mid-nineteenth century, making it one of the world's first energy provinces. By 1900 the Baku region produced about half the world's total crude oil. Since the early 1950s, however, several developments contributed to a substantial reduction of Caspian oil production. Concern over Baku's vulnerability to attacks during the Second World War, along with the discovery of oil in the Volga–Urals region of Russia, and later in western Siberia, led to a switch in the former Soviet Union's investment priorities. This new policy resulted in decreased exploration and production in the Caspian for most of the second half of the twentieth century. Since the late 1980s, however, Azerbaijan, Kazakhstan, and Turkmenistan have gradually occupied centre stage in the global energy markets. The three countries have succeeded in attracting massive foreign investment to their oil and gas sectors.

Since the collapse of the Soviet Union several international oil companies have negotiated and signed agreements with Caspian states, particularly Kazakhstan and Azerbaijan. These agreements suggest that the geological potential of the Caspian region as a major source of oil and gas is not in doubt. The rate of investment, however, is (and will continue to be) determined by the perceived risk in the region, or what industry experts call 'above-the-ground risk'. In other words, the risk is not in finding hydrocarbon deposits, but in juggling the multitude of risks associated with operating in very difficult host country environments. This section will examine the lack of consensus on the legal status of the Caspian Sea and the disagreement over the most cost-effective pipeline routes.

The legal status of the Caspian Sea

In the twentieth century the FSU and Iran signed several agreements to govern their relationship with respect to the Caspian Sea, most notably the Friendship Treaty of 1921 and the Treaty of Commerce and Navigation of 1940. Moscow and Tehran agreed that the Caspian was only open to their own vessels and was closed to the rest of the world. They also reserved a 12-mile zone along their respective coasts for exclusive fishing rights. No attempt was made to delimit any official sea boundary between them and the treaties said nothing about the development of mineral deposits under the seabed. Thus, many analysts and policymakers have questioned the applicability of these two documents to the new post-Soviet situation in the Caspian. Indeed, Russia, Iran, and the three former Soviet Republics have intensely disagreed on how to define the Caspian as a body of water.

A fundamental question in this debate on the legal status of the Caspian is whether it is a 'sea' or a 'lake'. According to the United Nations Convention on the Law of the Sea, nations bordering a sea may claim 12 miles from shore as their territorial waters and beyond that a 200-mile Exclusive Economic Zone (EEZ). If the law of the sea convention was applied to the Caspian, full maritime boundaries of the five littoral states bordering it would be established based upon an equidistant division of the sea and undersea resources into national sectors. If the law was not applied, the Caspian and its resources would be developed jointly, a division referred to as the condominium approach. After more than a decade since the break-up of the Soviet Union, the five littoral states have not agreed on whether to characterize the Caspian as a sea or a lake. The main point of contention centres around the uneven distribution of potential oil and natural gas riches in the basin.

The Russian position has varied over time. Initially, Moscow argued that the law of the sea did not apply to the Caspian because it was an enclosed body of water, and that regional treaties signed in 1921 and 1940 between Iran and the former Soviet Union remain valid. However, the signing of several agreements between the other three littoral states and international oil companies to explore and develop hydrocarbon resources beneath the Caspian's water prompted Russia to change its position. Thus, in 1996 Moscow proposed that within a 45-mile coastal zone each country could exercise exclusive and sovereign rights over the seabed mineral resources. Since the late 1990s, the Russian leaders have advocated the principle of dividing the seabed and its resources between neighbouring states. In line with this approach, Russia signed agreements with Kazakhstan (1998) and Azerbaijan (2001) dividing the northern Caspian seabed.

Unlike Russia, Iran has been more consistent in rejecting any bilateral agreement to divide the Caspian. Tehran's preference is for all five littoral states to adopt a collective approach in developing the mineral resources beneath the Caspian. Indeed, for the last several years, Iran has increasingly become the lone voice in the debate over the legal status of the basin. The reason is simple – Iranian shores on the Caspian seem to hold less oil and natural gas reserves than the other four littoral states. Since the break-up of the Soviet Union in 1991, the evolving positions of Azerbaijan, Kazakhstan, and Turkmenistan regarding the legal status of the Caspian have been driven by three interrelated developments. First, the coastal areas of each of the three countries are believed to hold more oil and gas reserves than those of Russia and Iran. Second, developing available hydrocarbon resources is considered crucial to the economic survival of these newly independent states, which have very few other economic assets. Third, the substantial international investments in the energy sectors of these three countries have incited them to be more assertive in their demands to divide the Caspian Sea into national sectors.

To sum up, the five littoral states have yet to agree on the legal status of the Caspian Sea. Despite this lack of consensus, a de-facto regime is emerging. Several international oil and gas companies have decided not to wait for an agreement and started developing the Caspian offshore fields. These ambitious and very expensive deals between international companies and littoral governments, however, face another serious hurdle – the lack of adequate system to ship the region's oil and gas to global markets.

Pipeline diplomacy

Given that Azerbaijan, Kazakhstan, and Turkmenistan are landlocked, they have to ship their oil and natural gas by pipelines that cross multiple international boundaries. The issue of potential routes through neighbouring countries has become a priority for both regional and international powers, as well as for oil companies. The construction of a pipeline would provide the transit states with several financial and political benefits, including access to oil or natural gas for their domestic needs, foreign investment and jobs, substantial transit fees, and political leverage over the flow of oil and gas. Thus, the process of choosing and constructing pipeline routes is complicated and requires delicate negotiations with many parties. Until recently, the existing pipelines in the Caspian region were designed to link the FSU internally and were routed through Russia. Most of the Caspian's oil and gas shipments terminated in the Russian Black Sea port of Novorosiisk. Upon independence, there are political and security concerns as to whether these Caspian states should remain so dependent on Russia as their sole export outlet.

For several years a number of proposed routes have been under consideration. These include a pipeline to the north to Novorosiisk (completed in 2000), a second one to the east from Kazakhstan to China, a third one to the south-west through Afghanistan to Pakistan, a fourth one to the south across Iran, and finally a pipeline to the west from Baku, Azerbaijan to the Georgian port of Supsa on the Black Sea (became operational in April 1999), or the Turkish port of Ceyhan on the Mediterranean (became operational in 2005). For several years international companies and the concerned governments have been engaged in serious negotiations to determine the priority of each pipeline. Both strategic considerations and financial interests have shaped the outcome of these negotiations.

Since the late 1990s, the United States has promoted the pipeline from Baku to Tbilisi to Turkey's eastern Mediterranean oil terminal at Ceyhan (BTC) as the main export pipeline (MEP). The first section of the pipeline was inaugurated in May 2005. Most of the oil comes from the Azeri–Chirag and Gunashli field complex in the Azeri sector of the Caspian Sea, but Kazakhstan intends to export some of its oil through this scheme. All the participants are expected to benefit from the BTC. Both Georgia and Turkey will earn transit fees and the BTC will reduce shipments through the Turkish straits of Bosphorus and Dardanelles. Ankara will also increase its leverage in Central Asia and the Caspian Sea. Azerbaijan and Kazakhstan will have access to European markets and reduce their dependence on Russia. The United States will further consolidate its economic and strategic relations with the Caspian states and reduce the Russian and Iranian influence in the region. The BTC pipeline is expected to be coupled later with a natural gas pipeline linking Baku and Tbilisi to Erzurun in Turkey's eastern Anatolia.

China has special interest in Kazakhstan's hydrocarbon resources. The two countries share long borders and Kazakhstan has the Caspian Sea's largest recoverable oil reserves, while its production is more than double that of Azerbaijan and Turkmenistan together. In recent years China has sought to increase its oil imports from Kazakhstan. Chinese policy reflects both Beijing's fast-growing needs for foreign oil supplies and its dissatisfaction with Russia's lack of commitment on a pipeline to ship Russia's oil to China. Thus, in May 2004 the two nations signed a joint declaration of what was termed the 'second section' of an oil pipeline project. The underlying rationale for this project is obvious. Kazakhstan intends to increase its oil production and ship it through multiple routes. Meanwhile, China needs to import a large volume of oil to maintain its impressive economic performance. Construction of this pipeline began in late September 2004.

Three conclusions can be drawn from this discussion of pipeline diplomacy in the Caspian Sea. First, given the domestic, regional, and international rivalries surrounding oil and gas fields in the Caspian, there is no doubt that multiple export routes would increase the energy security for consumers, producers, and the global energy markets by making deliveries less vulnerable to technical or political disruptions on any individual route. Still, energy security will have to be balanced by economic feasibility, since a larger number of pipelines would mean smaller economies of scale. Second, the decision to choose the most appropriate route reflects a competition between strategic concerns and economic interests. Most pipelines are built by companies, not by governments. Ultimately, projects must stand on their own commercial merit and the economics of a project will dictate its success. In the long-term, pipelines that make economic sense are more likely to be built than those that do not. Third, the pipelines' capacity and availability will, to a large extent, influence the timing of oil and gas development in the Caspian region.

To sum up, oil and gas production growth in the Caspian continues to be strong but not as initially expected. In addition to legal hurdles and regional and international rivalries, there is a growing concern about domestic political and social stability. According to Transparency International, a global think-tank that examines and assesses corruption, Azerbaijan, Kazakhstan, and Turkmenistan are among the most corrupt countries in the world.[3] This widespread corruption is coupled with the prospects of changes in the leadership. In October 2003 Ilham Aliyev succeeded his father Heydar Aliyev. The socio-economic and political upheavals in nearby Kyrgyzstan, Uzbekistan, Georgia, and Ukraine are likely to be echoed in the Caspian Sea region. This political uncertainty negatively impacts foreign investment. While the investment climate varies between the Caspian states, all of them are still in very difficult transition. Accordingly, the Caspian Sea region cannot be seen as a strategic alternative to oil and gas supplies from OPEC, particularly the Persian Gulf.

The Persian Gulf

Currently energy interdependence between OPEC producers, particularly the Persian Gulf states, and Pacific Asia is strong and is projected to grow further in the next few decades. Pacific Asia's consumption of oil and natural gas is projected to grow faster than any other region in the world as Tables 7.3 and 7.4 show.

On the other side of the energy equation, the Persian Gulf region holds the largest oil and natural gas reserves, cheapest cost-production, direct access to global markets, overall well-developed energy infrastructure, and spare capacity. Indeed, the Gulf's share in global oil production is much lower than its share of the world's oil reserves. In other words, the Persian Gulf region is under-exploited while most other regions, particularly Russia, the North Sea, and the United States, are over-exploited. Thus, it is widely projected that oil production from the Gulf will rise and the world will become more dependent on oil supplies from the region. Pacific Asia consumers already import most of their oil needs

Table 7.3 Oil consumption (2001–2025) in million barrels per day

County/region	2001	2025	Average annual percentage change
Australia/New Zealand	1.0	1.7	2.2
China	5.0	12.8	4.0
India	2.1	5.3	3.9
Japan	5.4	5.8	0.3
South Korea	2.1	2.9	1.3
Asia	15.6	28.5	2.3
Western Europe	14.0	15.7	0.5
USA	19.6	28.3	1.5

Source: United States Energy Information Administration. (2004). *International energy outlook*. Washington, DC: United States Government Printing Office. p. 167.

from the Gulf region. China and India seek to diversify their energy mix by relying more on natural gas and less on oil. Early exploration activities in the Persian Gulf had mostly concentrated on the search for crude oil rather than natural gas. Gas was considered a not-too-welcome by-product of crude oil production and was not utilized. This attitude has drastically changed since the mid-1970s. All Gulf producers have embarked on huge investments to develop their natural gas deposits. The Persian Gulf region is projected to play a significant role in global gas markets in the near future.

   Given these geological characteristics, Pacific Asia consumers have sought to further consolidate their energy partnership with the Persian Gulf producers. In 1999 the Chinese President Jiang Zemin visited Saudi Arabia and announced the creation of a 'strategic oil partnership' between the two nations. In 2004 China's Sinopec, along with other international oil companies, signed an agreement to explore for natural gas in Saudi Arabia. In the same year Sinopec signed a memorandum of understanding to buy 250 million tons of LNG from Iran over 30

Table 7.4 Natural gas consumption (2001–2025) in trillion cubic feet

County/region	2001	2025	Average annual percentage change
Australia/New Zealand	1.1	1.8	2.2
China	1.0	5.0	6.9
India	0.8	2.5	4.8
Japan	2.8	4.2	1.6
South Korea	0.7	1.8	3.9
Asia	6.4	15.3	3.9
Western Europe	14.8	23.7	2.0
USA	22.6	31.4	1.4

Source: United States Energy Information Administration. (2004). *International energy outlook*. Washington, DC: United States Government Printing Office. p. 168.

years. Iran will also export 150,000 b/d of crude oil to China after Sinopec develops the Yadavaran field. Since 1994 Iran, India, and Pakistan have negotiated the construction of a pipeline to export natural gas from Iran to the two large Asian consumers. Japan also is involved in several oil schemes in the Gulf, particularly in Iran and Kuwait.

Energy security: lessons and prospects

Several conclusions can be drawn from the discussion of energy outlooks in consuming and producing regions. First, the notion that energy security can be improved by reducing dependence on one particular region is unrealistic and misguided. The market for oil (and to a less extent for natural gas) is global and well-integrated. Second, given its geological characteristics, the Persian Gulf region will continue to be the driving force in the global oil market and in ensuring energy security. Third, Pacific Asia's growing energy needs suggests that its close energy ties with and dependence on the Persian Gulf will further grow in the foreseeable future. Fourth, this growing energy interdependence between the two regions is likely to have political and strategic ramifications. Historically, nations with great energy demand have sought to secure their energy resources by forging close political and military ties with their oil and gas suppliers. Strategically, Asian powers such as China and India share a similar stand with Persian Gulf states on issues such as the legitimacy of using nuclear energy for civilian purposes, opposition to economic sanctions, and the peaceful resolution of the Arab–Israeli conflict that would guarantee Palestinian rights. Fifth, there has been growing speculation that China's growing dependence on energy supplies from the Persian Gulf might prompt Beijing to expand its naval power and a rivalry between the United States (the largest oil importer) and China (the second largest oil importer) over the Persian Gulf (the largest oil exporter) might start.

This chapter argues against the last conclusion. The thrust of this study is that energy security should not be seen in zero-sum terms. Rather, continued dialogue and mutual understanding of common interests between consumers and producers will offer the appropriate conditions to establish and consolidate energy security. Within this context international organizations such as International Energy Forum (IEF) can play an important role in facilitating this cooperation.[4]

Notes

1 For a detailed discussion of the history and activities of these four companies see Wu and Han (2005: 18–25).
2 Attempts were made in 1966 and 1983 but ended in violence and military coups.
3 For further details see the organization website at www.transparency.org.
4 The IEF is an informal international forum to promote dialogue between the major players (i.e. consumers, producers, and oil companies) in the field of energy. The goal is to discuss common concerns, exchange information and policy views, and reach consensus on future challenges with an awareness of long-term common interests. The organization was established in Saudi Arabia in December 2003.

8 The impact of the new Asia–Pacific energy competition on Russia and the Central Asian states

Barry Naughten*

The prospect of increased exports of oil and gas from Russia and Central Asia[1] (henceforth R&CA) to the Asia–Pacific region rests on three important trends.

First, on the demand side, is a shifting of the centre of gravity from Europe, and to a less extent from the US, towards the rapidly developing economies of Asia. Of these, China is now the world's second largest oil consumer and its gas requirements are also rapidly growing. In contrast, requirements of the world's third largest oil consumer, Japan, are expected to remain static.

The second trend is increasing *import* dependence as domestic supplies of oil and gas are depleted in the United States, Western Europe and China – each currently 50–60 per cent import dependent with respect to oil. According to the United States Department of Energy (USDOE)'s International Energy Outlook (EIA, 2005b)[2] all three will be around 70 per cent dependent on oil imports by 2025, China's position having been transformed from a net exporter (22 per cent) in 1990 to a net importer (75 per cent) by 2025. International trade in gas, in both pipelined and LNG forms, is also expected to increase markedly.

The third trend is a projected tendency for increasing global dependence on oil sourced from OPEC and especially from the Persian Gulf states. However, such a trend will be markedly retarded to the extent that higher oil prices and other factors encourage increased oil production in non-OPEC economies and also restrain growth in total oil consumption.[3] A third offsetting effect would be 'supply diversification' policies of importing states – especially among those now heavily dependent on Middle East oil. Such offsetting trends and importer policies will favour the key non-OPEC economies of Russia and the Central Asian states directly bordering the expanding oil markets of the Asia–Pacific and standing to gain both economic and security benefits therefrom.

The R&CA states are adjacent to the Asia–Pacific markets and in the case of Russia and Kazakhstan actually share long borders with China, but very long distances are involved. Where sources and markets are close to sea-ports, it is by tanker that oil is most economically transported over such long distances. Pipelines are favoured where it is necessary to transport large volumes of oil, usually over shorter distances, to or from inland locations. Yet the oil and gas resources of Russia and the land-locked republics of Central Asia are typically far from ports and in some cases prohibitively so, such that the term 'stranded resource' (Foss, 2005: 113) can be applicable. Russia already has the world's most extensive oil and gas pipeline grid, and at least until recently, these Russian grids have also been the only conduit for oil and gas exported from the Central Asian Republics.

The focus of this chapter is less on the importers' perspectives, motivations and policy choices. Rather, it is on the hydrocarbon-rich R&CA producers standing to gain from access to these Asian markets and the extent to which such markets can contribute both to the socio-economic development of the R&CA states and their security. The chapter is in three parts. In the first section, the record and Asian market prospects of the oil and gas sector in R&CA are reviewed in terms of the available data and key official forecasts. R&CA export markets have hitherto been overwhelmingly directed westward to Europe. The second section deals with how oil and gas exports to the Asia–Pacific can contribute to growth of the R&CA economies. The background is, first, the exceptionally strong recovery of these economies since 1998, based in large part on the boom in oil and gas export revenues. However, there is a serious question about whether such growth can be sustained without major policy reforms, not least in the oil and gas sector. Such reforms have been embodied in the 'Putin model' that has had major implications for the oil sector and alleged abuses by so-called in the oil industry 'oligarchs' heading many of its private or bank-owned oil companies. Also considered are several politico-economic dysfunctions or 'diseases' arising from and potentially undermining resource-based economic growth. The third section addresses links between oil resource development and security issues. For example, are there benefits to the exporting states in greater 'demand diversity' than international free markets would provide? To what extent may an oil 'scramble' by OECD Asia and Developing Asia benefit or pose problems for R&CA states? What have been the responses of both the oil-exporting and oil-importing states of these two regions to the recent penetration of the US super-power into Central Asia – especially given its evident interest in that region's oil resources, as an alternative or supplement to global supplies from the Middle East? Growing influence of rival great powers or super-powers in a region that was previously a Russian preserve has led to the use of terms such as the 'the new great game' (Kleveman, 2004). These aspects of 'external stability' are linked to 'internal security' challenges arising within these typically autocratic states from opposition forces conveniently labelled, and perhaps sometimes opportunistically so, as 'radical Islamists'.

R&CA oil & gas: realising Asian market opportunities

Recent history of R&CA oil and gas production and export

A sharp fall in Russian annual oil production was evident from an all-time peak in 1987 to the late 1990s, corresponding to the economic collapse following the demise of the USSR and its centrally planned economy. This collapse meant a much reduced domestic requirement for oil. But far from a switch to exports

Table 8.1 R&CA: trends in oil supply and disposal (Mb/d)

	Production				Consumption				Exports			
	1989	*1992*	*1998*	*2004*	*1989*	*1992*	*1998*	*2004*	*1989*	*1992*	*1998*	*2004*
Russian Federation	11.1	8.0	6.2	9.3	5.1	4.5	2.5	2.6	6.1	3.5	3.7	6.7
Central Asia	1.0	1.0	1.1	2.0	0.9	0.8	0.5	0.5	0.1	0.1	0.6	1.5
Kazakhstan	0.5	0.5	0.5	1.3	0.4	0.4	0.2	0.2	0.2	0.1	0.4	1.1
R&CA	12.1	9.0	7.3	11.3	5.9	5.3	3.0	3.1	6.2	3.7	4.3	8.2
FSU	12.3	9.1	7.4	11.4	8.3	7.0	3.6	3.7	4.0	2.2	3.8	7.7

Mb/d = million barrels a day; R&CA = Russia & Central Asia; FSU = Former Soviet Union.
The lower level of exports from the FSU compared with R&CA was mainly due to oil transferred to
states such as Ukraine then part of the USSR but not of R&CA.

Source: BP–Amoco (2005).

occurring during this period, instead there was an equally calamitous fall in oil
export volumes and revenues. This was due in part to decline in world oil consumption and price after the Gulf War in the early 1990s but more to obstacles on
the Russian oil supply side arising out of both the excesses of the later Soviet
period (when oil supply capabilities were pushed to an unsustainable limit) and
chaos in the early years of transformation to a market economy.

A sharp export-led revival came with recovery from a serious collapse in the
international price of oil that was associated with the Asian financial crisis in
1997. Indeed, there has been debate about cause and effect of this collapse.[4]
Whatever the causal sequence, the recovery in Russian oil production and exports
was stimulated both by recovery in the world oil price in US dollars and also the
sharp decline in the rouble following the Russian financial crisis of 1998. The
recovery from this macro-economic crisis, both at the national and international
level, placed demand pressure on Russian oil supplies. By 2004, production levels were largely restored or even exceeded in the case of exports. Prospects for
Caspian oil have also improved, especially for Kazakhstan (Neff, 2005a, 2005b).

Table 8.2 R&CA gas production (TcM)

	Production			Consumption			Exports		
	1992	*1998*	*2004*	*1992*	*1998*	*2004*	*1992*	*1998*	*2004*
Russian Federation	597	551	589	417	365	402	180	187	187
Central Asia	111	76	134	72	70	85	39	6	48
Turkmenistan	56	12*	55	9	10*	16	47	2*	39
Uzbekistan	40	51	56	37	47	49	3	4	7
R&CA	708	627	723	489	435	488	219	193	235

R&CA = Russia &Central Asia; TcM = trillion cubic metres.

* 1998 was an anomalous year in which satisfactory commercial arrangements with Russia to take
Turkmenistan gas were lacking (Kuru, 2002).

Source: BP–Amoco statistics (2005).

Corresponding to its holding the world's largest gas reserves, Russia was also the largest producer (589 trillion cubic metres (TcM)) and exporter (187 TcM) in 2004. Uzbekistan and Turkmenistan are also significant producers. Turkmenistan (with much smaller domestic requirements) is also a significant exporter. In Russia, domestic consumption is subsidised such that domestic prices are $28/'000 cM versus $130/'000 cM for European markets.

However, the USDOE (EIA, 2005g) notes that both the Russian Government and the monopoly producer, Gazprom 'both project steep declines in natural gas output between 2004 and 2020'. But this refers mainly to three existing super giant fields. Russian liquefied natural gas (LNG) and pipelined gas exports to Asia are envisaged from Sakhalin (LNG especially to Japan) and pipelines from Irkutsk to Daqing (China) and from Kovykta fields to China and thence South Korea.[5] Major gas pipeline proposals are under consideration from Turkmenistan to China, Pakistan and Russia.

Projection of R&CA oil and gas for Asian and Pacific markets to 2020

A set of major pipeline projects for transporting R&CA oil and gas to Asian and Pacific markets over the period to 2020 is described in later sections. According to timetables now foreshadowed, completion of these projects could result in a pattern of trade for oil such as summarised in Table 8.3 comparing 2020 outcomes with those of 2004. Table 8.3 indicates that levels of oil exports to the Asia–Pacific region in 2004 were minimal, consisting mainly of oil imported by rail at considerable cost from West Siberia[6] to Daqing, China. However, if major pipeline projects proceed according to a plausible timetable, oil exports could be over 4 Mb/d by 2020, or ten times current levels. These exports could be divided relatively evenly between China and Japan as import markets, and also evenly between R&CA as sources.

Currently, both China's and Japan's oil imports are overwhelmingly from the Middle East. However, on the assumptions made, by 2020 the share of Japan's oil consumption – and imports – from Russia might reach over 30 per cent, with the Middle East contribution falling from 80 per cent in 2004 to 50 per cent. Similarly, the R&CA share of China's oil consumption could approximate 24 per cent. However, in China's case, the Middle East share of consumption is also projected to increase in absolute terms by a factor of 2.5 and its share of consumption from 19 per cent (37 per cent of current Chinese oil imports) in 2004 to 30 per cent of consumption in 2020. This reflects both the rapid increase projected for China's oil consumption (5 per cent annually compared with a slight *reduction* in Japan's case) together with a strong increase in China's import dependence (from 40 per cent to 70 per cent) reflecting also depletion of China's own domestic oil supplies.

Thus, the Chinese case exemplifies the projected global trend, even under the most favourable assumptions about the increasing supply role of R&CA, toward resurgence of the Middle East as key global supplier (Bahgat, 2005). Just as oil imported from R&CA would bulk large in Asia–Pacific oil consumption on these

Table 8.3 R&CA oil for Asian and Pacific markets: 2004 and 2020 (proj.)

Mb/d	2004*					2020†				
To: From	Japan	China	OA-P‡	Total exports	Total supplies	Japan	China	OA-P‡	Total exports	Total supplies
Russia	0.0	0.4	0.1	6.7	9.3	1.6	1.0	0.0	7.8	11.1
Central Asia	0.0	0.0	0.0	1.5	2.0	0.0	1.8	1.0	4.5	5.3
R&CA	0.0	0.4	0.1	8.2	11.3	1.6	2.8	1.0	12.3	16.4
Middle East	4.2	1.3	7.2	19.6	24.6	2.7	2.9	10.9	23.7	32.3
Other imports	1.0	1.7	2.0	20.3	12.2	1.0	3.1	3.3	27.7	15.0
Total imports	5.2	3.4	9.3	48.1	48.1	5.3	8.8	15.2	63.7	63.7
Domestic supply	0.1	3.1	2.2		32.7	0.1	3.5	3.7		46.4
Total consumption	5.3	6.7	11.5		80.8	5.4	12.3	18.9		110.1

Shares of total consumption (per cent)

	Japan	China	OA-P‡		Total supplies	Japan	China	OA-P‡		Total supplies
R&CA	0	6	1		14	30	23	5		15
Middle East	79	19	63		30	50	23	58		29
Total imports	98	51	81		60	98	72	80		58

R&CA = Russia & Central Asia.
*BP–Amoco statistics (2005).
†2020 projection based on Reference Scenario from USDOE IEO 2005 (EIA, 2005a), Middle East share is the residual, 'other imports' share remaining the same as 2004
‡Other Asia–Pacific.

Table 8.4 Oil production and net exports, R&CA, proj. 2020 (Mb/d)*

	2020 level		Implied increase on 2004	
Supply scenario: Consumption scenario:	High price† Low growth	Low price‡ Reference	High price† Low growth	Low price‡ Reference
Production				
FSU	17.7	15.8	6.3	4.4
Russia	12.0	10.6	2.7	1.3
Caspian and other FSU	5.7	5.2	3.7	3.2
Net exports				
FSU	13.0	10.1	4.7	1.8
Russia	9.0	7.0	2.3	0.3
Caspian and other FSU	4.0	3.1	2.5	1.6

FSU = Former Soviet Union.
*Based on USDOE *IEO 2005* (EIA, 2005a).
†US$48/b, 2025.
‡US$21/b, 2025.

projections, the trade would also become important to the export economies. For example, oil exports to Asia–Pacific could approach 20 per cent of Russian oil production by 2020 (compared with 3 per cent now) and 50 per cent of Central Asian production.

The above scenario is based on an optimistic set of assumptions with respect to mainly pipeline-based export of R&CA oil to Asia. To set this projection in context it is instructive to consider scenarios recently published by the USDOE (EIA, 2005a). These cases are distinguished by extreme assumptions about trends in the price of oil over the projection period as well as alternative rates of consumption growth. The USDOE's high price case of US$48/barrel (/b) is favourable to high cost non-OPEC producers such as R&CA and also induces lower growth in global consumption. In this scenario, the implied increment to Russian net exports is 2.3 Mb/d. This increment is consistent with an optimistic view on exports to Asia–Pacific. However, under the assumed low-price case of US$21/b, production of oil in R&CA is reduced and the increase in total exports is only 1.6 Mb/d. Such estimates are not intended to be taken too literally, but indicate some implications of price uncertainty, a subject to be considered further, below.

Is there enough oil and gas in R&CA to supply Asian markets?

R&CA oil accounts for between 6 and 13 per cent of global reserves depending on whether proven reserves or the most inclusive definition of economically recoverable reserves is used. Of proven reserves, Russia's is in the range of 60–72 billion barrels. Of 'proved and probable' ('2P') reserves of 116 billion barrels, only 7 billion has so far been identified in the vast regions of East Siberia and the Russian Far East (RFE).[7] As to Central Asia, even in respect of proven reserves, the level for Kazakhstan is contentious, being in the wide range of 9–39 billion

Table 8.5 R&CA: oil reserves and reserves/output ratios, 2020

| | Reserves, as at 2004 | | Projected total annual output | | |
| | | | 2020 | Reserves to output ratio | |
	'P1'* bn b	'3P'† bn b	Y Mb/d	'1P'/Y years	'3P'/Y years
Russia	60–72	146	11	15–18	36
Central Asia‡	17–54	204	5	9–30	112
R&CA	77–126	350	16	13–22	60

R&CA = Russia & Central Asia.
*Proven; higher figure is from *BP–Amoco*; lower is from *Oil & Gas Journal*.
†Economically recoverable: proven, probable, possible.
‡Corresponding to the range of 9–39 for Kazakhstan.

barrels with the lower figure published by the *Oil & Gas Journal* (2005) and the higher estimate by *BP–Amoco* (2005).

Pessimistic views about Russian supply capability exist (Dienes, 2004, 2005; Lambert & Woollen, 2004; Ebel, 2005). A conventional, if over-simple, measure of supply sustainability is the ratio (R/X) of proven reserves (R) to rates of output (X), giving a result measured in years of remaining output at that hypothetical constant rate. In the case of Russia, at the projected 2020 output of 11 Mb/d, but using the present (2004) estimate of reserves, this 'mixed' indicator[8] is in the range of 15–18 years, depending on whether the *BP–Amoco* or *O&GJ* estimate of proven reserves is used. A more sustainable export prospect emerges if economically recoverable Russian reserves can be expanded in line with above estimates, say toward the 'proven, probable and possible' ('3P') figure of 146 billion barrels, which would double the R/X ratio to around 36 years.

Any limits specifically on the currently under-explored East Siberian and RFE reserves and output might not be a problem as long as the pipeline system can be fed with output from West Siberia, given that new pipelines are linked to the existing grid. However, some reconfiguration of pipelines and associated expenses might be involved if flow directions have to be reversed or capacity requires expansion.[9] The corresponding calculations for Central Asia are even more problematic given the much wider of range of estimates (17–54 billion barrels) for its proven reserves. At a projected output of 5 Mb/d as at 2020, the ratio is in the correspondingly wide range of 9–30 years, but is 60 years if the '3P' value is used for reserves.

Russia's proven reserves of natural gas are higher than those of any other state, being 27 per cent of the global total, while Central Asia (mainly Turkmenistan and Uzbekistan) is a further 5 per cent. In addition, Asia–Pacific gas requirements are as yet on a much lower scale than oil requirements, though projected to grow strongly. Hence, availability to Asia–Pacific of R&CA pipelined gas is much less in doubt, the central question being its price competitiveness relative to LNG supplies, especially from the Middle East. Table 8.6 indicates changes in estimated reserves of both gas and oil over the last 20 years. Proven reserves of the FSU

Table 8.6 R&CA: proven reserves to output ratios, oil and gas, 1984–2004

	Proven reserves 1984	Cumulative output	Proven reserves 2004	Annual output 2004	Ratio	Proven reserves 1984	Cumulative output	Proven reserves 2004	Annual output 2004		Ratio
	R1984	Y1985–2004	R2004	Y2004	R2004/Y2004	R1984	Y1985–2004	R2004	Y2004		R2004/Y2004
	TcM	TcM	TcM	TcM	years	bn b	bn b	bn b	bn b	Mb/d	Years
	Gas					*Oil*					
Russia	na	11.0	48.0	0.59	81	na	26	72	3.4	9.3	21
Central Asia	na	2.3	9.2*	0.14	66	na	5	48	0.7	2.0	65
R&CA	na	13.3	57.2	0.73	78	na	31	120	4.1	11.3	29
FSU	37.5	13.8	58.5	0.74	79	81	31	121	4.2	11.4	29
Middle East	27.4	3.1	72.8	0.28	260	431	81	734	9.0	24.6	82
Asia–Pacific	7.0	4.2	14.2	0.32	44	38	28	41	2.9	7.9	14
Total world	96.4	43.3	179.5	2.69	67	762	269	1189	29.3	80.3	41

R&CA = Russia & Central Asia; FSU = Former Soviet Union.
*According to BP–Amoco statistics, in Central Asia, Kazakhstan (at 3 TcM) has the highest level of proven gas reserves then Turkmenistan (2.9) Uzbekistan (1.9) and Azerbaijan (1.4); the *Oil & Gas Journal* estimate is less, with Turkmenistan, Kazakhstan and Uzbekistan (each around 2 TcM) and Azerbaijan (0.9)M adding to 6.6 Mb/d.
Source: BP–Amoco (2005).

have substantially increased despite cumulative output in the intervening period, more or less on a par with the global experience for both oil and gas.

Investment in oil exploration and development

Existing estimates of proven and even '3P' reserves thus might give rise to concerns about the longer term availability of oil to meet the more ambitious export obligations. Is sufficient investment occurring in exploration and development of new fields? There are important distinctions to be made here. The first is between overall investment by the Russian oil companies and that component in exploration and development of new fields. A recent OECD Report fails to make this distinction.[10] Bank-owned private sector oil companies like Yukos were no doubt largely responsible for feeding the oil supply boom since 1998. However, this expansion in current rates of production was to the almost total neglect of their long-term investment in exploration and development of new fields (Gaddy, 2004: 349).

A second argument concerns the explanations for this imbalance. One is that property rights and the rule of law had been insufficiently secure to promote sufficient private investment in exploration as distinct from maximising output from existing sources. For example, Gaddy (2004: 350) blames 'the lack of a proper and developed market structure in Russia, in particular a high degree of uncertainty about property rights'.[11] The situation is more problematic than is suggested by Gaddy. This controversy will be taken up in the next section in the context of the 'Putin reforms'. The IEA (2004: 303) presents a less pessimistic picture on the development of new wells:

> The total number of wells in operation is now close to the number in 1990, despite a number of recent well closures. Average well productivity has rebounded, from 51 barrels per day in 1996 to 66 b/d in mid-2003, though it remains far lower than most other producing countries. ... higher prices have led to surge in investment and made possible partnerships with international and oil service companies. At 7.7 billion, total capital expenditure is more than three times higher than in 1999. Most investment is going to West Siberia, much of it to boosting output at already operating fields.

The levels of investment in exploration and development of new fields will be sensitive to expectations about future oil prices. As already noted, the non-OPEC R&CA countries are classified as high-cost producers and hence requiring higher prices for viability. IEA (2004: 304) notes that investment costs in E&D in new projects in East Siberia will be

> considerably higher than for existing brown-field projects in West Siberia because of a lack of infrastructure and more difficult geological conditions. The average investment needed per barrel of [daily] capacity stands at around $13,000, which is higher than in most parts of the world.

Assuming a risk-inclusive required rate of return of 15 per cent, this equates to $6/b. The average given by IEA (2004c) for 'transition economies' generally is $11,000 ($5/b), for OECD generally also around $13,000, but for the Middle East is $4,500 or only $2/b.

R&CA's modes of international oil transport and cost of pipelines

Some of the major oil pipelines under consideration are listed in Table 8.7. Why these pipeline projects in general and why these particular projects? This is a question that has to be answered on a case-by-case basis. Possible explanations include (i) economic; (ii) geopolitical security; and (iii) ecological.

In terms of economic explanations, relevant considerations firstly include cost relative to alternative modes. For example, more than 80 per cent of exported Russian oil was via pipelines in 2003, either cross-border or to sea terminals (OECD, 2004). Unit cost of pipeline transport to China is said to be approximately 20 per cent that of the stop-gap rail transport alternative. Currently, 0.2 Mb/d of oil (rising to 0.7 Mb/d) is exported to China by rail 2003–2010 in the absence of a pipeline. Hence, on these estimates, a pipeline appears highly commercial. Second, the unit pipeline costs in Table 8.7 may be compared with the future long-run average f.o.b. price of oil (say, US$20–48/b), or with production costs (US$11–13/b). A third comparison is with the so-called 'Asia premium'. According to Jaffe and Soligo (2004: 1), this latter concept

> refers to the fact that under the pricing formulae set by Saudi Aramco, the f.o.b. price for deliveries to Asian markets is set higher than for deliveries to U.S. and European markets … Ogawa (2003) states that the premium …

Table 8.7 Selected oil pipelines: R&CA to Asia–Pacific

Pipelines	Capacity		Length			Capital cost		
	Lower Mb/d	Upper Mb/d	miles	km	Total US$bn	per km $m/km	Unit cost‡ $/barrel
Russia							
Taishet–Nakhodka	0.6	1.6	2,298	3,698	12	3.2	3.5–9.0
					15	4.1	4.4–12.0
Angarsk–Daqing*	1.0	1.0	1,404	2,260	4	1.8	1.9
Central Asia							
Kazakhstan–Pakistan†	1.0	1.0	1,040	1,674	2.5	1.5	1.2
Kazakhstan–China:							
Aktyubinsk–Xinjiang	0.4	0.8	1,800	2,897	3.0	1.0	1.7–3.5
Atasu–Alashankou	0.2	1.0	614	988	0.7	0.7	0.3–1.6

*Now abandoned, to be possibly replaced with a spur line from the Taishet–Nakhodka line.
†Gwadar, Pakistan.
‡Assumes an interest rate of 15 per cent and a capital charge rate of 17 per cent; upper and lower bounds reflect the stated capacity (utilisation) range.

averaged 94 cents/barrel over the period 1991 – the first half of 2002 and has risen to over $1.50/barrel in some recent periods.

However, the proposed Angarsk–Daqing pipeline, available at a unit cost of around $2/b has been rejected in favour of the longer and much more costly Taishet–Nakhodka pipeline at $3.5–4.5/b at full capacity. An economic argument for Taishet–Nakhodka might be Russia's preference for competitive markets for the oil output at the seaport of Nakhodka, whereas the lack of alternative markets at Daqing suggests a monopsony on the part of China or at best a bilateral bargaining situation. Management and maintenance would also be simplified for a pipeline entirely in Russian territory. Non-economic arguments have also been suggested. One such refers to undue ecological costs associated with the Daqing route, especially in the sensitive region to the south of Lake Baikal. Finally, there are possible 'security' arguments in favour of the Taishet–Nakhodka route that will be considered in a later section.

Sufficiency of investment in the mega-pipelines: a purely 'economic' issue?

Particular risks are attached to large, capital-intensive projects, especially those lumpy projects with lengthy lead-times typified by long oil and gas pipelines.[12] The reality and perception of such risks had been brought into relief by the experience of the oil price shock of the 1970s. That shock led to quite durable expectations of future oil prices that were far in excess of actual price levels. These expectations led (*inter alia*) to over-investment, for example, in oil-saving energy supply technologies such as coal-fired and nuclear-powered electricity generation, that were large, lumpy, capital-intensive, and with long construction lead times. A result was costly under-utilisation of such assets and a disinclination to repeat this experience, especially given the continued long-term uncertainty in oil prices.

Thus, investment risk may be particularly great when the unit value of output – here the price of oil – is highly uncertain and dependent on future macroeconomic conditions and policy contexts that are also uncertain. The relevant output price is that ruling when the projects are finally commissioned and averaged (and discounted) over the subsequent lifetimes of those projects. The modern theory of investment (Pindyck, 1991; Dixit, 1992) suggests that rational decision-making under such circumstances typically entails 'high' real required rates of return, say equal to 2–3 times the risk-free rate.[13] This is why a rate such as 15 per cent has been used in calculating unit costs in the above table. Under these circumstances, it will often make sense to defer investments until more information is available, or to shorten lead-times, where this is possible and cost-effective.

The oil price spike of 2004–2005 has reached levels in real terms unprecedented since the 1970s. Long-term uncertainty in the price of oil remains extreme, if we are to take seriously the USDOE's published alternative crude oil price paths over the projection period to 2025, being in the range from US$21/b to US$48/b.[14] Such future price uncertainty means that high risk-inclusive rates of return would apply to private investment in capital-intensive crude oil projects involving long lead-times, including pipelines. This suggests a rational basis for the history of deferral of such decisions by both private investors and the nation-states involved.

Thus, where pipeline mega-projects have been under consideration for very long periods but not implemented, it is not obvious that such deferrals are due to 'bureaucratic' delays supposedly characteristic of the state or statist methods, as sometimes asserted – for example, by Lahn and Paik (2005). Deferral of such projects pending more reliable information about the future may be rational in many circumstances, whether the leading role is taken by a private or a government agent. Even where expected consumption requirements press strongly, it may be reasonable to persist with more reversible, modular technologies or the use of existing infrastructure even where the unit cost of these options may be high – even the temporary use of costly rail transport of Russian oil to China.[15] However, the use of costly rail is most typically motivated by a pattern of bottlenecks in the existing pipeline grid. This latter phenomenon exists in conjunction with under-utilisation in other parts of the grid, in both cases seen as a legacy of Transneft's monopoly dating from the Soviet era (IEA, 2004: 305–306).

The existence of what appear to be generous financial commitments (co-ordinated) by the Japanese government with respect to the Taishet–Nakhodka pipeline suggests that state intervention may be accepted as necessary.[16] However, a question arises on the extent to which such a policy position, with implied subsidy, reflects a rational *economic* logic as distinct from a wider logic of *security*. Currently (as noted above), Japan obtains a large share of its oil imports from the Middle East and there is debate about the existence of an 'Asia premium' on the CIF price of this crude oil in Japan and Asia generally (Jaffe & Soligo, 2004: 3; Ogawa, 2004). Part of the rationale for a pipeline alternative may be to do with exercising leverage to reduce this premium. Answering this question would start with a comparison of this premium (estimated at around US$1/b) with any (subsidy component of) the unit cost of oil transport via pipeline. It will be recalled that calculations of the unit cost of pipelines suggested as much as $4–5/b for Taishet Nakhodka when fully operational at 1.6 Mb/d, but considerably lesser amounts for other proposed international oil pipelines. It is not clear how these costs will be shared between the producer and importer, for example, the extent of any subsidy that might be involved in Japan's substantial financing of this project.[17]

Sakhalin: an oil and gas project not tied to long pipelines

The oil and gas project proceeding in the Sakhalin peninsular of the RFE[18] will supply 0.34 Mb/d of oil from Stages I & II by 2006 plus LNG for Japanese electricity generation. Ownership of Stages I and II of this project includes a strong Japanese component. Bradshaw (2003: 83–84) notes that Sakhalin is remarkable for its progress to date compared with other projects for Russian oil and gas west of the Urals. He attributes this in part to Sakhalin's favourable location adjacent to Asia–Pacific markets and hence the available option of LNG rather than pipelines for the gas output, arguing that:

While LNG projects required substantial investment in liquefaction and re-gasification, they are flexible in that an LNG buyer can choose between numbers of suppliers. Traditionally, the LNG trade has been based on long-term contracts, but there is increasing evidence of the emergence of a spot market of sorts. Pipelines, however, are far more inflexible. They require substantial reserves at one end to sustain and fill the pipeline and a significant market at the other to justify the investment. Once built they cannot be moved and lock the seller and the buyer into a long-term relationship. They are also relatively fragile (as evidenced by recent problems in Iraq). Pipelines that cross more than one international boundary are even more complicated as by definition they require multilateral cooperation.

Asian oil and gas markets: an engine for R&CA development?

The GDP growth experiences of R&CA states since the fall of the USSR were characterised by economic collapse followed by strong recovery in respect of both the oil and gas sectors and these economies at large. These two phases are starkly evident in Table 8.8. They were marked off by the Russian financial crisis of 1998, the subsequent depreciation of the rouble and the recovery of world oil prices from a deep trough in late 1998. The positive links between the fortunes of the oil sectors and the economies as a whole are clearest for Russia and for major oil and gas exporters of Central Asia, but growth in the second phase was also significantly stronger for other Central Asia states.

President Vladimir Putin's objective, announced in his address to the State Duma May 16 2003 is to 'double GDP in the next ten years', that is, to sustain the 7 per cent annual growth actually experienced since 1998. To put this goal into perspective, it should be noted that the Russian share of world GDP is now

Table 8.8 GDP, population (2004) and trends since 1992 and 1998

	GDP US$					Population	
	'000/ capita	bn, PPP	% pa growth since		–	%pa million	since
	2004	2004	1992– 2004	1992– 1998	2004	2004	1992
Russia	10.2	1449.2	2.3	−3.9	8.9	142.4	1.1
Azerbaijan	4.0	33.1	3.0	−5.4	12.0	8.3	−1.1
Kazakhstan	7.4	111.3	3.6	−3.4	11.1	15.0	1.1
Kyrgyz Republic	1.9	9.9	2.1	−2.1	6.5	5.1	−0.4
Tajikistan	1.2	7.9	2.1	−5.8	10.7	6.3	1.1
Turkmenistan	7.3	35.9	5.6	−6.2	18.8	4.9	2.1
Uzbekistan	1.8	45.8	3.3	1.6	5.0	25.9	1.6
Central Asia	3.7	243.9	3.6	−2.9	10.5	65.6	0.8
R&CA	8.1	1693.0	2.5	−3.7	9.1	208.0	0.0

R&CA = Russia & Central Asia.

Source: IMF (2005).

only 1.7, a long way from economic super-power status. Indeed, it has been suggested that Mexico and Brazil (similar in per capita GDP terms) are appropriate comparisons (Shleifer & Treisman, 2004).

Oil sector as stimulus to Russian growth since 1998

An OECD Report on the Russian economy (2004) provides a fairly detailed assessment of the links between the oil sector and its GDP growth since 1998. Of obvious importance was the strong recovery in world oil prices from their deep trough in 1998 and also the stimulus given to oil production mainly from intensive production from existing wells. As early as 2000, the oil and gas sector was back to contributing as much as 24 per cent of Russian GDP, comparable with other major oil exporting states such as 22 per cent in Norway and 32 per cent in Iran (IEA, 2004: 289).

The sustainability of this mechanism depends on the future path of international oil prices and its impact on exploration and production investment and levels. It also depends on avoiding certain well-known pathologies[19] of rapid, volatile or uncertain growth influenced by resource export revenues. Ideally, oil and gas production should be managed so as to broaden the economic base and cause least disruption, for example, through fiscal policy: profits and rent taxes, and equalisation funds to compensate for price and export revenue fluctuations.

In the OECD's assessment (2004: 17) increased production by private oil companies had been a key factor in Russian GDP and export growth.[20] However, this wholly favourable OECD view of the role of the private oil companies in Russian economic growth is challenged in President Vladimir Putin's model for restructuring the oil sector.

The 'Putin model' and oil industry restructuring

The 'Putin reforms' are wide-ranging and can be viewed in at least two ways that may not necessarily be totally mutually exclusive. The first interpretation is as primarily a 'political' defence of his position and that of an autonomous centralised state, given that the head of Yukos, Mikhail Khodorkovsky and other oil oligarchs (Goldman, 2004) had been positioning themselves to strengthen their political control using their oil funds. This enhanced their commercial position under a regime that would allow strong links with international oil companies (IOCs). The economic corollary of this assessment of the reform program is that of reversion to a statist central planning model reflecting supposed limitations of Putin's background as an ex-KGB bureaucrat.

In the second interpretation, the reforms are steps towards an economic regime that would still be market-driven but would require private sector agents to respect and not subvert a strong state role in industrial and energy policies. This resurgent state role is not reversion to a discredited pre-1992 economic model but represents a rational response to failures of the 'neo-liberal' or 'Washington Consensus' imposed post-1992, and to unsustainable (mis-)management of the private oil sector.

Critics of Putin's reforms (implicitly Gaddy, 2004; OECD, 2004) argue on the contrary that whatever the abuses, the growth record since 1998 suggests that such a private sector approach was warranted. However, Boussena and Locatelli (2005) argue persuasively for the second of these positions, that economic growth is unlikely to be sustainable in the absence of Putin's reforms. Although their major focus has been on the oil industry, part of his government's position is that over-dependence on the oil industry as major basis for economic growth is itself unsustainable (Johnson's Russia List, 2005).[21]

The reforms have included the break-up of Yukos, the largest private oil company, the charging and gaoling of Khodorkovsky and others on charges including tax evasion, and merger of Yukos' main operational arm with state-owned Rosneft.[22] The OECD concluded that the private oil companies had a key role in the recovery of Russian oil industry, and hence of its GDP. This need not be denied, but it can be argued that this mechanism is not sustainable. The problem is not merely the private sector's lack of confidence in the stability of property rights and the rule of law but that its own 'oligarchic' and corrupt behaviour threatened these desirable conditions. The alternative was extension of joint ventures with private IOCs, moving toward the 'transparency' supposedly habitual to these latter companies. But this was regarded as equally a threat to this core sector's acting in the national interest. In terms of state ownership, the resemblance to the 'OPEC revolution' of the 1970s is unmistakable, recalling that few of these states had any sort of a heritage in central planning and state ownership. However, it is still the case that much of the Russian oil sector remains in private hands and with a significant foreign component.

Boussena and Locatelli argue that the 'Washington Consensus' prescription or 'big bang' model for introducing a neo-liberal capitalism in Russia had brought a strong rent-seeking bias on the part of 'stake-holders' and inordinate (potential) political power. Results of this bias included:

- short-term maximisation of share values by maximising current revenues and share values;
- buying and selling of companies at the expense of long-term investment in productive assets, wasteful production practices and under-spending on exploration and the development of new fields;
- asset stripping;
- weak rule of law;
- undermining the business transparency required for a properly functioning market regime;
- tax avoidance and evasion: both preconditions and effects of the strong rent-seeking bias.

The reforms require strong state action to:

- increase exploration expenditure on new oil fields albeit jointly with IOCs or other sources of high technology as necessary;

- optimise taxation of oil and gas rents for the wider socio-economic benefit;
- retain state control over national output levels of oil to the extent necessary for co-operation with other global producers, especially OPEC, to obtain price discipline.

In addition, Russian foreign policy could not be pre-empted by private sector agents ('freelancers') such as Yukos, independently of government, pursuing international projects such as the now defunct Angarsk–Daqing oil pipeline. The Taishet–Nakhodka pipeline favoured by the Government is now apparently to proceed with strong involvement of Government enterprise: that is, Rosneft in production (ironically, now including the former Yukos subsidiary, Yuganskneftegaz), with the pipeline monopoly Transneft responsible for constructing and operating the pipeline.

Central Asia: economic development through oil and gas?

As noted above, the Central Asian republics vary considerably in their dependence on the oil and gas sector, with Kazakhstan dependent on increasing oil exports and Turkmenistan on expanding gas exports.

Dysfunctionality of the 'rentier state'

The rentier economy has been defined as one 'in which income from rent dominates the distribution of national income, and thus where rentiers wield considerable political influence' (Bromley, 1994: 94). As such, it is the basis for a 'rentier state', one that is 'reliant not on extraction of the domestic population's surplus production but on externally generated revenues, or rents, such as those derived from oil' (Anderson, 1990: 61). In a country with a state enriched by taxes or charges on resource extraction and export only low rates of personal income tax need be imposed.[23] A political consequence is that citizens who would otherwise be paying taxes can more 'legitimately' be denied political rights, because a key element of the social contract is absent. This denial can be sought where the ruling élite perceives the modernisation process accompanying industrialised growth as a threat to its narrow interests and political power, where the 'modernisation process' includes higher education,[24] economic production, consequent development of human and social capital, and of civil society.

Sapamurad Niyazov's Turkmenistan has been described by the International Crisis Group (ICG, 2004) as 'one of the world's most repressive regimes'. Turkmenistan is the largest Central Asian gas exporter and has a relatively small population. Hence, its GDP/capita (US$7,300) is moderately high by Central Asian standards, approaching that of Russia (US$10,200). Resource-based funds are thus available for attaining domestic stability partly through this rentier state model (Kuru, 2002). Certainly, the proportion of Turkmenistan's labour force in 'industry' (as distinct from agriculture and 'services') is very low (14 per cent) and its unemployment rate extremely high (60 per cent). Turkmenistan's ruling élite has also sought to avoid disruptive change through an insulated and neutral foreign policy, failing for example to join in Central Asian regional organisations. However, it has been far from isolationist in its efforts to establish a capacity to export its gas by pipeline in ways that are independent of the monopsonist power of Russia and its grid system, including both Iran options and (preferred by the US) routes through Afghanistan.

With a much lower GDP/capita (US$1,800), Uzbekistan produces similar amounts of gas. But a much larger population implies (i) that much less gas is available for export due to greater domestic requirements; and (ii) greater diffi-culty in co-opting popular support with this smaller amount of export revenue. Its other staple commodity had been cotton. Hence, Uzbekistan conforms less closely to the model of a rentier economy or state and more to one of simple repressive autocracy.

Even those Central Asian states less well endowed with oil or gas wealth, have benefited from indirect economic effects of the post-1998 oil boom. Anticipating the security focus of the next section, it is possible to see Russia's oil-based eco-nomic growth as enabling its deployment of 'soft power' with respect to its Commonwealth of Independent States (CIS) neighbours and the Central Asian Republics in particular (Hill, 2004: 18–21). A large part of this 'soft power' is a strengthened economic relationship, for example, through Russia's ability to pro-vide consumer goods and jobs for guest workers and hence large income trans-fers to their home economies in the form of remittances. Similarly, China's presence and security interests in Central Asia have been bolstered by a broadly based commercial interpenetration with these economies.

Diversifying Central Asian oil & gas exports from Russian pipeline monopsonist

When they were components of the USSR, all the oil – and gas-producing republics of Central Asia had been totally dependent on the Russian oil and gas grids for dispersal to markets and lacked the option of autonomous export. Post-1992, independence brought their right to engage independently in international trade. But for these land-locked states there was no physical option in the short-term but continued dependence on the Russian grid, and this at a time when Russia was itself undergoing the economic stresses of transition.

As will be considered further in the next section, with the downfall of the Soviet Union, the United States had both vastly increased apparent international power and also the 'security' incentives to encourage independence efforts in Central Asia. Indeed, supporting this vision may have been part of the explana-tion (or rationale) for the myth of the Caspian as the 'new Saudi Arabia' in terms of its oil-bearing potential.

The Baku–Tbilisi–Ceyhan (BTC) oil pipeline was opened early 2005. For at least two geopolitical reasons, this consortium-owned pipeline was strongly sup-ported by the United States (SEEN, 2004). First, it completely by-passes both the existing Russian grid system and its territory, offering independent access to

Western oil markets not only to Azerbaijan but potentially also to Kazakhstan (Gorst, 2004). Second, it represents an alternative to a new pipeline through the 'rogue state' Iran (Chanlett–Avery, 2005: 18).

As well as through BTC, Kazakhstan is well positioned to take significantly greater advantage of its untapped oil wealth also through significant direct pipelined exports to China. With these options and enhanced bargaining strength, its access to the Russian oil grid will be on potentially more favourable terms, especially given the 'soft power' turn of Russia itself, as just noted. That turn, with its commercial bias, is in large part a consequence of Russia's own oil-driven economic prosperity since 1998.

Energy security and interdependence in the R&CA/Asia–Pacific relationship

Oil 'supply diversity' and a shift from the Middle East source to R&CA

The above discussion of the major oil pipeline proposals, and especially major ones such as Taishet–Nakhodka, referred to justification project-by-project against commercial criteria bearing in mind data such as freight costs versus costs of the pipelines and theoretical constructs such as the 'Asia premium'.

Another set of explanations for the persistent strong interest in such projects is to do with national security, and 'energy security' in particular. This includes questions such as the vulnerability of sea routes for Middle East oil, especially the 'choke points' of Hormuz and Malacca and their policing. It involves the possible benefits of supply diversity in improbable and extreme situations, such as war and trade sanctions. Increasing Asian unease about the Middle East oil source cannot be separated from the worrying political prospects of that region, especially since the deteriorating situation in Iraq post-2003.

The security aspects of the various land-based pipeline options themselves require case-by-case consideration. Some proposals entail transit through only one national territory (for example, Russia in the case of Taishet–Nakhodka), while others cross several national territories including unstable sub-national regions with insurrectionary agendas – one example, among many, being a proposed gas pipeline from Turkmenistan–Afghanistan–Pakistan–India (Misra, 2005).

Conversely, opening an oil or gas pipeline may offer a decisive purpose or pretext for central Governments to exert increased control over potentially unstable or strategically important remote regions within their own territory. An example, again, may be the Taishet–Nakhodka pipeline, which penetrates deep into remote East Siberian and the RFE territories that might become problematic if little economic activity occurred in those regions.[25] In a geopolitical sense, there may be Russian concerns about Chinese or even Korean encroachments in these remote regions.

Putin has demonstrated his sensitivity to the issue by recently making regional governorships a central government appointment. Such regional development arguments may conflict with market-driven imperatives or purely economic efficiency criteria, for example, as embodied in the Washington Consensus. Gaddy and Hill (2003), for example, point to potentially high costs of development policies in such harsh and remote regions, including high costs of the oil developments and pipelines themselves.

The Taishet–Nakhodka project, together with the development of the oil and gas resources of Sakhalin, also binds together two nation-states that have a major unresolved security dispute about the Kuril Islands since World War II, itself not unrelated to regional oil and gas prospects.[26] This 'binding together' would be reinforced not only by the prospectively high share of Russian oil in Japan's consumption (30 per cent by 2020) but by Japan's initiative in financing the bulk of the pipeline project.

In a geopolitical perspective, the Russian choice of Japan rather than China as a willing development partner in this project, could also be seen as part of a 'balancing' posture against the US super-power, not only in the Middle East but in North Asia where the United States' asymmetric alliance with Japan is a dominant consideration. However, from the Russian perspective one attractive feature of the Nakhodka destination is again that it is a seaport and as such makes available tanker-borne Russian oil not only for the Japanese market but precisely also to the US – as well as China and other North Asian markets – perhaps complementing other projects to supply Russian oil and gas to the US, for example, from Murmansk. Similarly, proposals exist for the piping of Sakhalin oil and gas to China as well as the main intended markets in Japan. In each case, the essential fungibility of oil (and gas) is demonstrated, and the attractions and costs of bilateral arrangements may appear beside the point.[27]

China's remote Xinjiang province may be a second relevant example of pipelines as an adjunct to regional development having a security objective in its own right. Historically part of Turkistan and with a majority Muslim population, Xinjiang is also vital transit territory for proposed oil pipelines from Kazakhstan (Atasu to Ala-shankou and Aktyu-binsk to Xinjiang province) and a gas pipeline to China from Turkmenistan.

Japan's history of policies favouring energy supply diversity goes back at least to the oil crises of the 1970s when its oil companies' overseas investment activities were subsidised to that end. Today, apart from Russian oil and gas from Taishet–Nakhodka and Sakhalin, Japan's oil companies have a presence in Azerbaijan and Kazakhstan (Uyama, 2003) as well as the Azadegan field in Iran (Chanlett–Avery, 2005).

A similar 'supply diversity' theme is evident in China's actions and those of its oil companies. As well as the pipelines from Sakhalin, alternative spur lines to Daqing from the Taishet–Nakhodka pipeline remain on the agenda.[28] Associated with its oil pipeline proposals from Kazakhstan, PRC's state-owned oil company CNPC has acquired a 60 per cent stake in the Kazakh state firm Aktobemunaigaz (Chanlett–Avery, 2005: 9) along lines of its bilateral and commercial deals in the Middle East (especially Iran) and East Africa[29] (Klare, 2004: 161–175; Xu Yi-Chong, 2005; Zweig & Bi Jianhai, 2005).

Supply security inherent in a pipeline grid depends on it being sufficiently complex and inter-connected so as to allow the flexibility of re-routing in the event of temporary outages of particular links due to accident or sabotage. This complexity and increasing inter-connectedness is potentially a positive 'network externality' from each incremental pipeline link. However, a regulatory regime (multilateral if necessary) may need to be in place to guarantee operational access. To the extent that such an enhanced grid functions effectively, the oil and gas transported takes on more of the character of a fungible commodity, like sea-borne oil. This contrasts with a bilateral monopoly (or monopsony) situation inherent in a single international pipeline between two inland points, such as the defunct oil pipeline proposal from Angarsk (East Siberia) to Daqing, China.

The Middle East and Central Asian presence of the US super-power

China's and Japan's increased interest in diversifying away from Middle East oil is clearly related to perceptions of instability in this region, and to some trepidation about the role of the US therein. This concern is especially pressing in the case of China, its potential Asia–Pacific regional rival. The US influence in the Middle East has been portrayed as a (benignly) hegemonic and stabilising one. For instance, Walter Russell Mead has characterised this role as follows:

> We (that is, the U.S.) do not get that large a percentage of our oil from the Middle East. Japan gets a lot more … And one of the reasons that we are sort of assuming this role of policeman of the Middle East, more or less, has more to do with making Japan and some other countries feel that their oil flow is assured … so that they don't then feel more need to create a great power, armed forces, and security doctrine, and you don't start getting a lot of great powers with conflicting interests sending their militaries all over the world.

This passage is cited by Schwarz and Layne (2002: 36) who, as IR 'defensive realists', are critics of current neo-conservative or nationalist US foreign policy. Mead's formulation can be read in different ways. For example, it is not easily distinguished from 'Carter Doctrine' as framed in 1980 by Zbigniew Brzezinski with respect to direct US control of the Gulf region rather than it being delegated or shared, as it was in the preceding era of 'twin pillars' (of Saudi Arabia and the Shah's Iran) of Nixon Doctrine.[30]

However, by 2005, China and other Asia–Pacific powers might be expected to have reservations about the notion that this US security role is either necessarily stabilising or an unambiguous 'global public good' as formulated in hegemonic stability theory. One concern has been the destabilising effects of the US war on Iraq (2003–) and its subsequent occupation, including with respect to Iraq's oil industry.[31] The notion of diversifying the Asia–Pacific's oil supply sources to include R&CA has no doubt been reinforced by growing concerns about the reliability of a US-managed or US-stabilised Gulf as the world's prime oil supplier. However, such concerns have not precluded Japan and China from actively cultivating stronger bilateral ties, both commercial and arms-related, with particular Gulf oil sources such as Saudi Arabia and Iran.

US competence to perform its long-standing role of stabiliser, or guarantor of the status quo, through techniques such as containment or deterrence, consistent with the above 'Mead formulation', may thus be in question. However, a distinct source of disquiet lies in the new elements in 'Bush Doctrine' (Bush, 2002) underpinning the Iraq engagement and including dimensions such as 'preventive war', militarily imposed 'regime change' and negativity toward multilateralism in favour of a radical unilateralism (Jervis, 2005).

A parallel concern arises from US moves toward establishing a position in Central Asia in the wake of the USSR's collapse, and more recently under the justification of the post-'9/11', anti-Taliban war in Afghanistan. Is the US strategy of extending its hegemony beyond the Middle East to the Central Asian region to be seen as an extension of the 'status quo' logic expressed in Mead's formulation? Or is it the beginning of a 'transformative' process in line with Bush Doctrine? However, in August 2005, the US had to accept Uzbekistan's eviction of its base in that country, established to pursue its global 'war on terror' (Marquardt & Wolfe, 2005; Weinstein, 2005a, 2005b).

Russia's previously noted 'soft power' stance in Central Asia is in the interests of its revised 'near abroad' policies (Hill, 2004e, 2004f). One objective is no doubt to offset this deepening US penetration of Central Asia. The stance is also part of both a rivalry and 'balancing' rapprochement with China, in the context of R&CA meeting a major share of China's future oil requirements. Finally it is a matter of stabilising regional relations.

Initiated by China and partly motivated by similar fears, is the 'Shanghai Cooperation Organization' (SCO). The SCO includes Russia, Kazakhstan, Uzbekistan and two of the other Central Asian Republics. Not included is gas-rich Turkmenistan which, as noted above, has a well-entrenched stance of isolationist neutrality, seeking gains by playing Russia, United States and China (ICG, 2004) off against each other. Weinstein (2005) noted that 'India, along with Pakistan and Iran, sought and was granted observer status in the S.C.O., an acknowledgment of the organization's growing geostrategic importance .... As the S.C.O. grows in strength, Washington's influence in Central Asia will diminish'.[32]

Even if the US perceives an organisation such as the SCO as a form of balancing against itself, some forms of such 'balancing' might be regarded as less threatening than others. For example, Chinese (or Japanese) links with R&CA may be more acceptable to the US than links, especially militarily, with Middle East states such as Iran. Others would argue that the SCO has more to do with reinforcing stable regional relations than with balancing against the US as such – that is, doing away with the need for a global hegemonic 'policeman'. As a precedent, even that might not be viewed benignly by some strategists of U.S. hegemony.

'Competitive' aspects of energy security policy

The notion of 'competition' is problematic. In neo-classical economic theory,[33] 'competition' is regarded in a benign or 'positive sum' sense as a mechanism leading to efficient allocation of scarce resources, and as such is preferred to 'collusion' or the market power of a monopolist. Competition may for example include a process of excess demand bidding up the price of oil in the face of supply constraints. The rising price has benefits in suppressing over-consumption and efficiently rationing available supplies, while encouraging investment in expanded future supplies and the introduction of substitutes for oil, including its efficient conservation. An implicit assumption is that property rights are firmly established and transparent in the context of the rule of law.

This is not to say that all developments surrounding the market are benign. Recent global supply constraints have been blamed on rising uncertainty, especially about future oil prices, due to a range of factors. That high-risk premia in required rates of return suppress investment has already been noted. Given the consequent constraints on supply, strong growth in oil consumption, especially from China, has been a prime cause of the current 'super-spike' in the price of oil – an increase of over 80 per cent in the 18 months to August 2005 (Harris & Naughten, 2005). In the simplest economic meaning, 'access' refers to oil being freely available on the international market as a tradable, fungible commodity. Such market access is not normally in question under peacetime conditions, though policies such as tariffs and quotas can change the terms of that access.

Broader notions of 'access' encompass not just the energy commodities but also the flow of productive capital associated with exploration for and development of energy. IOCs play key roles here along with state-owned national oil companies (NOCs). Increasingly, the latter are also operating internationally, but especially in ways that involve bilateral state-to-state relations. Rather than leaving oil to be distributed through free (spot) markets or absorbed by vertically integrated IOCs, resources may be 'tied up' or 'locked in' on a bilateral basis, often under long-term treaties and/or contracts. The bilaterally negotiated pipeline projects fit this description. However, commitments of large amounts of finance and resources to such infrastructure may require state-guaranteed markets and prices. In other cases, states may restrict the flow of capital inflow through IOCs or foreign NOCs. As noted above, such restrictions are now a marked trend in Putin's Russia and have characterised the Gulf and OPEC states since the 1970s.

In this more political context that includes supply (and demand) security, 'competition' thus has a wider meaning. The decision-making agents are nation-states rather than firms and the context is the (global) system of states, rather than simply the market. The so-called 'resource scramble'[34] refers to the sense of competitive urgency on the part of the oil-importing states in this process of capital-export to secure future oil supplies.

In this broader context, the notion of 'competition' may take on more of a 'zero sum' connotation and the assumption of clearly defined property right might not hold. Oil-dependent states and their associated corporations can come into potential conflict requiring resolution: diplomatic, through international law or otherwise. In the more extreme cases, oil-related disputes may involve actual ownership and territorial conflicts, for example, in the South and East China Seas (The Economist, 2005), the Spratleys, the Kuril Islands (all in the Asia–Pacific but involving Russia and Japan in the latter case) as well as jurisdictional issues in the Caspian Sea involving not only the Central Asian, but also the littoral states of Russia and Iran.[35]

The less benign potential of Asian energy competition has been underlined in more alarmist terms by Calder (1996, 1997a, 1997b) or Luft (2004), while others such as Manning (2000b) and Jaffe (2001a) have adopted a more sanguine view that emphasises efficient markets, transparency, international law and effective diplomacy.

Demand side responses

The above responses to the problem of energy security are all on the supply-side. As such they can be distinguished from policy actions that involve restraining consumption of oil and gas – 'demand-side' policies. Such policies can reduce oil import dependence by reducing oil dependence in general. They include tightening of vehicle fuel efficiency standards, increasing rates of tax on transport fuels[36] and reducing rates of subsidy on domestic oil or gas use. In international comparative terms, among major Asia–Pacific economies, existing fuel tax rates tend to be much on the low side, especially in China and South-East Asia. The position is even more distorting than this since governments including those of China and Indonesia have also subsidised oil consumption especially in periods of high crude oil price, or kept product prices low by enforcing price controls on refiners (Petroleum Economist, 2005) resulting in supply shortfalls.

Policies of domestic demand restraint can be warranted as much in oil-exporting economies such as R&CA as in importing ones. In fact, the former (Middle East, Russia, Indonesia)[37] are prominent among states under-taxing or subsidising fuel use, thereby reducing the amounts of oil and gas available to earn export revenue, and accelerating depletion. Such policies can have at least three socially efficient purposes: (i) reducing external diseconomies associated with transport fuel use, such as traffic congestion and urban air pollution; (ii) relieving 'security dilemmas' associated with supply-side 'resource scrambles'; and (iii) countering the market power[38] of low-cost oil producers (Middle East and OPEC generally) likely to increase with its market share as is projected.

Unilateral policies of demand restraint can be motivated by the local and national domestic benefits of type (i). However, the important global or regional benefits (ii) and (iii) also warrant a multilateral approach that will enhance global demand restraint, thereby stabilising oil market outlooks and mitigating the interrelated problems of 'price super-spikes', long-run higher oil prices, and 'resource scrambles',

Conclusions

The addition of these new eastern markets for oil and gas to the existing western ones can be seen as enhancing the 'demand security' of R&CA. For Russia, there is the additional benefit of opening up the remote regions of Eastern Siberia and RFE and their resources. At the current level of Russian-proven oil reserves, its total output, which is approaching the historic peak of the 1980s, could be sustained for only a further 20 years. If projected exports to Asian markets are added and existing export and domestic markets are to be retained, this theoretical lifetime could fall significantly, lending urgency to expansion of proven reserves by increased investment in exploration and development, and putting the 'Putin reforms' to the test.

From the Asia–Pacific perspective, 'diversifying supply' can mean tying up supplies for long periods through bilateral arrangements. The resulting 'resource scrambles' could become threats to peace and stability. In purely economic terms, the associated foreign investments could also prove costly for the importing states such as China. Alternatively, they could be viewed as part of a learning process for an economy growing with great rapidity from a low base, and with excess capital for export that is not yet profitably deployable at home. But it is not always clear that the appropriate lessons are learned.

Some risks can be diminished by firmer expectations of continued well-functioning global oil and gas markets. However, this requires internationally secure property rights and there may be diminishing confidence among the rising importers of Asia that this task can be left to the US as the sole global and regional hegemon, for example, as regards the protection of sea-lanes or management of Gulf peace and stability.

Both security and economic benefits would be enhanced by regional arrangements that are as inclusive and multilateral as possible. Examples include the consultative management of equitable access to pipeline grid systems, two-way flows of capital and technology, joint security arrangements to defend infrastructure against sabotage. Such supply-side approaches should not preclude more effective demand-side management for energy, preferably occurring multilaterally in both importing and exporting economies.

Notes

* Useful comments were provided by Professor Stuart Harris and Dr Kirill Nourzhanov, both of CAIS, Australian National University. Glenda Naughten provided editorial assistance.
1 The Central Asia states are variously defined. Narrowly, they are often identified with the five states Kazakhstan, Kyrgyzstan, Tajikistan, Turkmenistan, and Uzbekistan all of which were part of the FSU, are to the east of the Caspian and are part of the much larger historic region known as Turkistan that also includes, for example, western provinces of China, known as the Xinjiang Uygur Autonomous Region and Afghanistan north of the Hindukush, see http://www.encyclopedia.com/html/T/

Turkistn.asp. As used here the term also includes the FSU state of Azerbaijan bordering the west Caspian. Other states of the region are important as actual or potential transit states for oil or gas pipelines from the region, for example, Georgia (of the FSU), Iran (also a Caspian state as well as a Gulf state) and Afghanistan. The term 'Commonwealth of Independent States' (CIS), also in use, includes several FSU states outside Central Asia.

2 That is, USDOE's Reference scenario: import ratios (per cent) relative to consumption (EIA, 2005a) being:

	1990	2002	2010	2025	
United States	43	53	56	66	Became a net importer in 1971
Western Europe	63	50	55	66	With North Sea's decline, imports are rising again
China	-22	42	60	75	Became a net importer in 1993

3 In the IEO 2005 Reference scenario (EIA, 2005a) in which oil price is projected at US\$35/b (2025), Middle East share of production increases from 24 per cent (2002) to 31 per cent (2025); the OPEC share from 40 per cent to 44 per cent.

4 Morse and Richard (2002) argued that the collapse in the international oil price was not entirely due to the Asian financial crisis of 1997–1998 and the associated fall in demand. In their view, it was also induced by Saudi Arabia's failure to persuade the Russian Federation (a non-member of OPEC) to curtail its increased supply during this period. Morse and Richard were led to speculate that Russia would prove an effective long-run 'balancer' against Saudi Arabia and OPEC in moderating oil prices. Their assessment of Russia's capabilities in this regard was strongly challenged by critics soon after (Telhami & Hill, 2002). These same sorts of sentiments no doubt had some part in exaggerated estimates of Central Asian reserves and production capabilities common amongst US official commentators in the 1990s.

5 Chanlet–Avery (2005: 13) noted that:
   the initial proposal for a \$17 billion dollar gas pipeline, running from Irkutsk through Beijing and under the Yellow Sea to South Korea, would have served the Chinese and South Korean markets. In June 2004, however, Japan emerged as a potential buyer, and Russian negotiators suggested an alternative pipeline that would parallel the proposed Angarsk–Nakhodka oil pipeline, and therefore serve the Japanese market as well

6 Watkins (2005) reports that the goal had been 10 million tonnes (0.2 Mb/d) by 2005 but these deliveries are falling short due to rail capacity constraints (2005 estimate based on first three-quarters).

	2003	incr.	2004	incr.	2005
million tonnes	3.6	68%	6.0	22%	7.6
Mb/d	0.07		0.12		0.15

7 A data difficulty compounded by increasing state control of Russian oil exploration is that reserves are regarded as a state secret, rather than being to a degree the target of corporate transparency rules.

8 There is no one 'correct' ratio here. The more usual ratio is that of current reserves divided by current output, but this ratio would be reduced if future output growth is taken into account. An estimate of reserves as at say 2020 would be extremely problematic but would need to exclude depletion during the period up to that date and include new discoveries and revised estimates, the latter by definition unpredictable.

9 See the proposed Adria pipeline reversal (EIA, 2005a):
   Reversal of the Adria pipeline, which extends between Croatia's port of Omisalj on the Adriatic Sea and Hungary (see map), has been under consideration since the

1990s. The pipeline, which was completed in 1974, was originally designed to load Middle Eastern oil at Omisalj, then pipe it northward to Yugoslavia and on to Hungary. However, given both the Adria pipeline's existing interconnection with the Russian system, and Russia's booming production, the pipeline's operators and transit states have since considered reversing the pipeline's flow, thus giving Russia a new export outlet on the Adriatic Sea.

10 In respect of the growth period 1998–2004, OECD (2004: 42) notes that:

Strikingly, the growth of oil-sector investment was led by companies controlled by the state or by oil industry insiders: by 2000, their investment was already 70 per cent above 1998 levels. By contrast, oil companies owned by major financial groups (whose owners' property rights were perceived as less secure) were investing only marginally more than in 1998 ... In 2001, however, as perceptions of the security of property rights further improved, the latter group of companies began rapidly increasing investment, soon reaching levels comparable with the former group. This investment led to a sharp increase in oil production and exports in the following years.

11 According to Gaddy (2004):

Some have argued that YUKOS' short-sightedness was a natural manifestation of its 'predatory' capitalist profit-seeking approach. In fact, a more credible explanation for YUKOS' failure to invest in development was the lack of a proper and developed market structure in Russia, in particular a high degree of uncertainty about property rights. In framing a strategy for their company, YUKOS' owners had to ask themselves the question: Why invest for the sake of future profits if we cannot be sure we can keep those profits? In contrast, the 'insiders' – the owners and managers of Surgutneftegaz and Lukoil – could feel more secure thanks to their special relationships with government officials.

(Gaddy, 2004: 350)

12 In the case of the Taishet–Nakhodka pipeline, construction lead-time is 3–4 years for throughput capacity of 0.6 Mb/d (2010); the infrastructure requirements for expansion to 1.6 Mb/d (2020) are not yet clear.

13 According to Dixit (1992):

... even when the cost of capital is as low as 5 percent per year, the value of waiting can quite easily lead to adjusted hurdle rates of 10–15 percent. Summers' (1987) finding of median hurdle rates of 15 percent is no longer a puzzle.

(Dixit, 1992: 117)

14 This low price is well above the actual trough occurring in 1997 which was less than US$10/b.

15 According to Blagov (2005):

However, the planned oil exports to China could hardly serve as a substitute for the Angarsk–Daqing pipeline project. Rail freight is expensive, and while Yukos was keen to export oil to China by rail, other Russian oil firms are understood to be reluctant to replace Yukos as supplier to China due to high costs and low profit margins

(Blagov, 2005)

16 According to Blagov (2005):

Tokyo has been lobbying for an oil pipeline route to the Pacific. To back up its lobbying, Japan reportedly promised up to $14 billion funding of the pipeline as well as $8 billion in investments in the Sakhalin-1 and Sakhalin-2 oil and gas projects, according to Russian media reports. The estimated cost of the oil pipeline from eastern Siberia to Nakhodka could reach $11–12 billion. The Taishet–Nakhodka route is seen as a strategic asset for Russia, allowing it to funnel crude not only to Japan but to Korea, Indonesia, Australia and the US west coast as well.

(Blagov, 2005)

17 A final (?) feasibility study of the project is under way at the time of writing.

18 Sakhalin as a whole (USDOE EIA, 2005a) is said to involve 14 billion barrels of oil
    and 2.7 TcM of gas (the latter being about the same as Turkmenistan). Stage I has a
    capital cost of $12 billion and expected output of 0.25 Mb/d oil and Stage II 15 billion
    and 0.09 Mb/d plus gas. Russian equity involvement appears to be in Stages I (?) and
    IV–VI but not Stage II, which involves 'two Japanese companies, Mitsui & Company,
    and the Mitsubishi Corporation, own a total of 45 percent of Sakhalin Energy.
    Ensuring the viability of a project that could cost $10 billion, four electricity and gas
    companies in Japan have signed long-term contracts to buy the liquefied natural gas
    from Russia' (Brooke, 2004). The other 55 per cent of Sakhalin II is owned by Royal-
    Dutch Shell. According to a press release from the project (Oilvoice, 2005):

    > The Sakhalin II development represents the largest foreign direct investment proj-
    > ect underway in Russia. It was the first Production Sharing Agreement (PSA) to be
    > signed in Russia and the first PSA to go into operation. ... Phase 1 has been pro-
    > ducing oil from the Vityaz Complex offshore Sakhalin since July 1999. ... Sakhalin
    > Energy has sold its crude oil to refineries in seven different major markets – Japan,
    > Korea, China, Taiwan, the Philippines, Thailand and the USA. ... Year-round oil
    > production is expected in 2006, and deliveries from the new LNG plant are planned
    > to commence end 2007.

19 The main examples are 'Dutch Disease', 'rentier economy', and 'resource curse'.
    Dutch disease refers to a situation in which expansion of resource exports elevates the
    real exchange rate potentially squeezing the remainder of the tradable sector, for
    example, manufacture of import-competing or exportable goods. Regarding its man-
    agement in Russia since 1998, see Vatansever (2005) and OECD (2004). The 'rentier
    economy' is discussed below in the context of Turkmenistan. The 'resource curse'
    refers to cases in which there is an apparent empirical link with stunted development,
    under-investment in human capital, poverty, political corruption and civil strife or war
    (Karl, 1997; Sachs and Warner, 1999; Ross, 2001, 2004a, 2004b).

20 According to the OECD (2004):

    > Since 2000, the importance of the private oil companies' performance for the econ-
    > omy as a whole has been enormous. Industry accounted for slightly below half of
    > GDP growth in 2000–03 and the oil sector for somewhat below half of industrial
    > growth. Since the state-owned companies barely grew, this means that Russia's pri-
    > vate oil companies directly accounted for somewhere between one fifth and one
    > quarter of GDP growth. Taking into account the knock-on effects from oil-sector
    > procurement and wages on domestic demand, the actual contribution of the private
    > oil companies to economic growth was probably greater still. Moreover, the private
    > oil companies have played a crucial role in keeping Russia's external balance in sur-
    > plus, and thus in allowing the current consumption boom to unfold.

21 In its Russia Country Analysis Brief, the USDOE (EIA, 2005f) comments as follows
    on Russia's 'dangerous dependence' on energy export revenues:

    > In 2004, Russia's real gross domestic product (GDP) grew by approximately 7.1%,
    > surpassing average growth rates in all other G8 countries, and marking the country's
    > sixth consecutive year of economic expansion. Russia's economic growth over the
    > last five years has been fueled primarily by energy exports, particularly given the
    > boom in Russian oil production and relatively high world oil prices during the
    > period. This type of growth has made the Russian economy dangerously dependent
    > on oil and natural gas exports, and especially vulnerable to fluctuations in world oil
    > prices. (*emphasis added*) Typically, a $1 per barrel change in oil prices will result
    > in a $1.4 billion change in Russian revenues in the same direction – a fact that
    > underlines the influence of oil on Russia's fiscal position and its vulnerability to oil
    > market volatility.

22 Yuganskneftegaz, with a production capacity of 1 Mb/d, became part of the state-
    owned Rosneft (0.5 Mb/d). With other state-owned companies being 0.7 Mb/d, total
    state-owned oil production now amounts to about 24 per cent of the total. A proposal

to merge Gazprom (the natural gas monopoly) Rosneft has been much discussed and negotiated but so far (September 2005) not implemented.

23 In the context of the political economy of the 'rentier state', one consequence of the oil boom in the Russian Federation itself is that income taxes have been cut to a flat rate of 13 per cent, among the lowest in the world (Hill, 2004: 16). Not only is this low reliance on income tax but its flat rate is consistent with more extreme variants of the Washington Consensus. At the same time, a major element of implied welfare transfer to the population is in the opaque (and non-discretionary) form of heavily subsidised gas prices.

24 Probably reflecting the USSR heritage, reported literacy rates in all the R&CA states are extremely high (97 per cent +) and with little gender bias, in marked regional distinction with, for example, Iran (86 per cent males; 73 per cent females) or especially, Afghanistan (51 per cent males; 21 per cent females); source: CIA fact-sheets.

25 The first planned stage of the Taishet–Nakhodka project is a 2,400-km oil pipeline from Taishet to Skovorodino. The second stage 'involves further construction from Skovorodino to Perevoznaya' and 'depends on development of East Siberian oil fields' (Watkins, 2005a).

26 Chanlett–Avery (2005: 13) notes:

Putin's inclination to accept the Japanese proposal is buttressed by strengthening economic relations between Japan and Russia. Bilateral trade grew by 25 per cent in 2003, fueled by the energy sector's growth in Sakhalin. Japanese investment in Russia also rose by nearly $1 billion between late 2002 and Spring 2004. As economic ties develop, Tokyo has made diplomatic overtures to Moscow, pledging to work towards resolution on the Northern Territories dispute dating from World War II, announcing bilateral ministerial visits, and urging more Russian involvement in Northeast Asian affairs.

27 'Fungibility' refers to its international tradability as a quasi-homogeneous commodity. However, there are qualifications to do with quality, location, refinery configuration, market characteristics and end-use environment standards. For oil, 'quality' has two main dimensions: sulphur content (or sour versus sweet) and hydrocarbons (heavy versus light). 'Location' refers to matching of supply sources to market characteristics so as to minimise freight and refining costs. For an accessible, comprehensive and up-to-date account, see Skinner (2005). In the event of mismatch or capacity shortage, importers also have the option of importing refined products. While oil's fungibility is associated with low transport costs but complicated in these ways, the fungibility of internationally traded gas (essentially methane) has been compromised by its higher transport cost, a deficiency that is disappearing as the LNG market develops.

To the extent that pipelined gas is less and LNG markets remain immature, concerns of importing states about costs of the supplier state's political leverage remain: for example, in Europe with respect to its growing dependence on Russian gas (Dempsey, 2004; Pravda RU, 2005). Again, a complex interconnected grid with alternative supply sources and governed by principles of open access will relieve such concerns.

28 Still keeping controversy alive in this tri-nation saga is the possibility that a Daqing 'spur' line could be completed before the full completion of the pipeline to Nakhodka.

29 Corden (2005) notes that, coincident with its economic boom, China has the highest investment ratio globally, but for now may have reached the limits of cost-effective investment domestically. Hence, unusually for a developing economy, there is an economic rationale for it being a major capital exporter. In the case of its FDI, if not its investment in US treasuries, this can be seen as partly a learning experience.

30 Brzezinski continues to insist on continued US dominance in 'Europe, the Far East and the Persian Gulf' (2004: 17). In turn, this line does not seem to differ greatly from the neo-conservative position that the US ought to maintain absolute and permanent dominance regionally as well as globally. The essential reference for neo-conservative policy is the Project for a New American Century (PNAC, 1997) and as expounded by

Lemann (2002) and Armstrong (2002). It became the new orthodoxy when embodied in the President's National Security Strategy Statement (NSSS) (Bush, 2002).

However, neo-conservatives are also set apart by other dimensions such as their objective of Middle East-wide 'regime change' as a key component of US foreign policy, the principle of preventive war and the hostility to multilateralism – aspects about which Brzezinski's support may be less enthusiastic.

Less extreme proponents of US global dominance emphasise notions of global leadership, legitimacy, hegemony and economic power as distinct from the sheer military dominance underlined by the neo-cons. Mead's notion also fits into this 'hegemony model' (in a Gramscian sense) as long as the (erstwhile) stabilising role of the US in the Gulf remains perceived as also in the interests of 'subordinates' such as Europe and Asia; alternatively theorized as the provision of 'global public goods' to these 'free riders'.

Schwarz and Layne's 'off-shore balancing' realist position (2002) is sharply distinguished from those of both the neo-conservatives and Brzezinski, and also the global hegemony theorists, when they argue:

> For more than fifty years American foreign policy has sought to prevent the emergence of other great powers – a strategy that has proved burdensome, futile, and increasingly risky. The United States will be more secure, and the world more stable, if America now chooses to pass the buck and allow other countries to take care of themselves.

How other countries – especially oil importers such as China and Japan – are to be 'allowed' to 'take care of themselves' remains to be seen, as is the independence allowed to oil exporters in such arrangements.

31   Iraq is the location of the world's second largest reserves of low-cost oil after Saudi Arabia but its peak output of 3.5 Mb/d occurred as long ago as 1979. In March 27 2003, then Deputy Secretary of Defense Paul Wolfowitz had said that Iraq's oil revenues could bring between US$50 and $100 billion within 2 or 3 years following the country's 'liberation'. 'We're dealing with a country that can really finance its own reconstruction, and relatively soon' (Luft, 2005). Yet two and half years after the occupation began (August 2005), Iraq's oil output was still only 1.8 Mb/d compared with optimistic projections, now abandoned, that it might reach 6 Mb/d by 2010.

32   Applying the general principle of non-interference specifically, the S.C.O. declaration at the end of the July 5 2005 summit called for a timetable to be set for the closure of US military bases in Uzbekistan and Kyrgyzstan that support Washington's operations in Afghanistan, but are also elements of Washington's strategy of creating a permanent arc of bases spanning East Africa and East Asia. Following the summit, the Uzbek Foreign Ministry issued a statement that the US Khanabad airbase could serve no other purpose than support operations for the Afghan intervention: 'Any other prospects for a US military presence in Uzbekistan were not considered by the Uzbek side. Washington responded that were Tashkent to insist on closure of the Khanabad base, the US had other options' (Weinstein, 2005).

33   Compare with Schumpeter's more combative conception, influenced by Marx, of capitalist competition. This is one in which 'market power', although shifting, is typical and not the exception. However, Schumpeter goes on to note, anticipating Baumol's notion of 'contestability' that the market 'disciplines before it attacks' (1943–1966: 85) so that agents with market power are themselves under continual threat from new entrants, especially, it might be added, in the case of global markets.

34   Perhaps the term's classic use was in Europe's 'scramble for Africa' of the post-1870 period.

35   According to the EIA's Caspian Country Analysis Brief (December 2004):

> Although there is still no overarching agreement between the five Caspian littoral states on the division of the Sea's resources, three states have come to a trilateral agreement on sub-surface boundaries and collective administration of the Sea's

waters. In May 2003, Russia, Azerbaijan, and Kazakhstan divided the northern 64 per cent of the Caspian Sea into three unequal parts using a median line principle, giving Kazakhstan 27 per cent, Russia 19 per cent, and Azerbaijan 18 per cent. Following this, development of the northern Caspian Sea's hydrocarbon potential, where most of the region's oil reserves and largest international projects are found, will likely move forward despite the lack of a comprehensive regional consensus. Meanwhile, offshore development in Turkmenistan and Iran, both of which refused to sign the May 2003 agreement, could fall even further behind.

36 Other policy measures include the adoption and progressive tightening of vehicle fuel efficiency regulations (as recently introduced in China) and the wider extension of such policies, for example, to include SUVs (exempt under the US CAFE system but to be included in the Chinese equivalent).

37 Steps to reduce oil subsidies are now being taken in Indonesia (washingtonpost.com, 2005). China has now adopted vehicle fuel efficiency standards effective from July 2005, stricter than those of the US and to be tightened in 2008 (USA Today, 2004) and recent reports suggest that 'Beijing is likely to consider a 20–50 percent tax on retail gasoline and diesel prices, which [currently] are among the world's lowest' (Chen Aizhu, 2005).

38 There is something of paradox here, or perhaps a self-correction mechanism. An effect of demand restraint brought about by investment in oil-saving technologies will be to reduce the price of oil, for example, as in the IEA's 'Alternative scenario' (2004: 367). In the medium term, this will tend to increase the market share of the low-cost producers in OPEC and correspondingly reduce production and attenuate depletion of oil from higher cost non-OPEC producers such as R&CA.

Part III

Implications

9 Strategic dimensions of energy competition in Asia

William T. Tow

As part of the world's most dynamic region for projected economic growth and geopolitical change, Asia–Pacific countries will become more import dependent for its energy demands and will require more comprehensive strategies to ensure their future access to energy resources. Among the key issues regarding this requirement are: (1) 'energy nationalism' versus regional and international market co-operation; (2) energy source diversification from fossil fuels to nuclear energy development and coal consumption, with implications for nuclear proliferation and environmental politics; and (3) the intensification of both contingent and structural risks to Asia–Pacific energy security (Koyama, 2001; NBR, 2004). After briefly discussing how 'geopolitics' relates to energy and resource issues in the Asia–Pacific, this chapter will assess each of these three issues. It will integrate each issue into broader considerations of strategic relations between and military capabilities of the region's key energy players (the United States, Japan, China and India) and into an evaluation of possible measures to modify energy competition in the Asia–Pacific's strategic context.

The paper's basic argument adapts a variant of Simon Bromley's model for understanding US energy geopolitics. Geopolitics emphasises the spatial dimension of international relations, including assessing polities who control or aspire to control those material assets that allow them to dominate bodies of land and water at the strategic expense of other polities. Rather than viewing geopolitics as shaped by commercial ventures seeking new energy reserves and markets or by rising concerns about levels of import dependence, it is argued here, one should instead ask how the broader nature of geopolitics shapes energy security policies (Bromley, 2005: 225). Geopolitical competition has been traditionally formed by the inter-relationship 'of trade, war and power, at the core of which were resources and maritime navigation' (Le Billon, 2004: 2). More recently, international relations theory (especially international political economics) has tended to gloss over resource scarcity and competition in favour of emphasising 'resources of economic value'. Neo-liberal institutionalism and globalization theories directly link the growth of capitalism and international trade to the efficiency of resource use and the inherent reduction of interstate 'resource wars' (Peters, 2004: 189–191).

Contemporary Asia–Pacific geopolitics, however, conforms more to the 'traditional' model that anticipates conflicts rising from resource scarcity. East Asia (largely spearheaded by China) and South Asia (mainly led by India) are experiencing an explosion in fossil fuel import demand. Between 1990 and 2003, Asia accounted for approximately 75 per cent of the world's total increased demand for oil consumption (increasing from approximately 8 million barrels per day (MBD) to just below 15 MBD (Herberg, 2005b). While producing just 11 per cent of the world's oil supply it is now consuming 21 per cent of it, importing about 44 per cent of its needs compared to only 7 per cent in the early 1970s (ADB, 2004). The region's import dependence will only grow over the next few decades. International Energy Agency projections estimate that petroleum imports will more than double (to 36 MBD) by the year 2030 and much of this supply will originate from the Persian Gulf. Asia now accounts for 75 per cent of the world's liquid natural gas (LNG) trade – a demand that will treble by 2030 and will be serviced by extra-regional supplies five times as great as they were in 2002 (especially from Russia and the Persian Gulf). Japan, Korea and Taiwan are 100 per cent dependent on LNG imports while China and India are heading toward upward of 50 per cent LNG dependence over the next few years (Herberg, 2005b).

China's growth as an energy consumer is notably spectacular. In 2003, it surpassed Japan to become the world's second largest oil consumer after the United States and is now the world's fifth largest importer of oil. More tellingly, China's energy consumption per unit of gross domestic product (GDP) is five times greater than the US and 12 times greater than Japan (Austin, 2005: x). While its current LNG usage is low (3 per cent of its overall energy consumption), that rate will more than double (to 8 per cent) by 2010. China is moving rapidly to replace coal consumption, which accounts for two-thirds of China's total energy consumption, with cleaner LNG supplies. Overall, as Mikkail E. Herberg has observed, these patterns of energy supply and consumption result in 'a profound and deepening sense of energy insecurity in Asia that promises to have important long-term geopolitical implications for the region …' (Herberg, 2004: 340).

Moreover, as asserted above, geopolitics is driving energy security behaviour in the region, not the other way around. New and previously unlikely state-centric alliances and coalitions are now being forged with energy calculations largely in mind. China is courting the Gulf Cooperation Council countries (especially Saudi Arabia, Oman and Yemen) more intensively, visibly enhancing its already key role in supporting Iran's oil industry and orchestrating economic and military *rapprochement* with Russia and various Central Asian states. It is courting Indonesia and Australia heavily to fulfil its future LNG needs, complicating US geopolitical calculations in peninsular South-East Asia and the South-West Pacific. Japan has likewise cemented its oil ties with Iran (particularly with a contract to develop the Azadegan oil field in partnership with Iran's national oil company) and has outbid China for the construction of a large oil pipeline from the East Siberian region of Angarsk to the Pacific coast of Nakhodka (Blagov, 2005).

South Korea and India have also worked to diversify their energy sources in the Persian Gulf, parts of Africa and Central Asia.[1]

Diversification of energy collaborators, and the introduction of massive infrastructure investments to solidify these new links, reflect a growing mercantilist competition that is no less pervasive or ruthless than its historical geopolitical counterparts fuelled by Europe's industrial revolution.[2] Contemporary resource planners, however, have failed to bargain as effectively as their European colonial predecessors on how to divide the spoils of the world's non-renewable energy supply. Resource competition between the Asia–Pacific's rising powers (China and India) and its more established ones (the United States, Russia and Japan) could emerge as a critical source of future regional tension. In this context, energy nationalism, energy source diversification and energy risk assessment warrant greater attention.

Energy nationalism

As Herberg has observed 'energy nationalism' is now becoming predominant in Asia: that is, direct competition between states to control regional energy supplies that leads to a closer integration between energy and strategic relations (Herberg, 2004: 368; Herberg, 2005). This is reflected in Sino–Japanese territorial disputes over offshore natural gas fields in the East China Sea and near the Senkaku/Diaouyu Islands. Russian–Japanese collaboration on energy development could potentially lead to a serious security dilemma emerging in North-East Asia, with China and both Koreas feeling increasingly marginalised or 'contained'. To be sure, institutional efforts by the Asia–Pacific Economic Cooperation (APEC) forum to co-ordinate emergency oil stockpiling in the region, and by the Three-Party Committee, including China, Japan and South Korea that has entered into formal discussions on energy security within the ARF framework, are encouraging (APERC, 2002; Cossa, 2003). But the challenges of economic nationalism will remain formidable in Asia as they continue to be linked with the fundamental national security concerns of the major players there. Over the short-term, three key challenges will accentuate this linkage: (1) de facto dependency on potential security rivals for the stability of energy supplies (especially China's dependence on US naval power); (2) a condition of incessant political security crises in many important energy production locales which accentuate already strained production infrastructures; and (3) the threat of serious interdiction of oil supplies at key transit points between those locales and Asian end-users.

Is China an American energy protectorate?

Although China has moved precipitously to establish energy resource footholds in the Middle East and Central Asia, it remains concerned that US global naval power could be applied to impose energy containment against itself in a future showdown with the United States over Taiwan or elsewhere in the region

(Downs, 2004: 31–32). It is unclear to what extent the current Chinese leadership accepts earlier arguments posited by Western strategists that China should enjoy 'free-riding' on the back of US military power that can secure the region's sea lanes and thus its energy supply access without commensurate Chinese naval capabilities.[3] The US Department of Defense has speculated in its latest annual assessment of Chinese military power (released in late July 2005) that Beijing's cultivation of special relationships with distant energy suppliers in the Middle East, Africa and Latin America could lead it towards increased investments in a blue-water capable fleet. At present, however, most of China's conventional military development appears targeted toward fighting and winning short-duration conflicts along its peripheries (Office of the Secretary of Defense, 2005: 10–13).

For the time being, however, China's maritime power remains focused on sea denial rather than on broader sea control and SLOC protection missions. This raises what Chinese president Hu Jiantao referred to in late 2003 as the 'Malacca Dilemma'. Eighty per cent of China's oil imports pass through the Strait of Malacca but Chinese surface combatants cannot project power there. Moreover, 'its limited organic defence capability leaves [its] surface ships vulnerable from attack from hostile air and naval forces' (Office of the Secretary of Defense, 2005: 33–34). Until it is able to project more long-range and sustainable air support capabilities to support its few destroyers and submarines capable of operating beyond the South China Sea, the security of China's oil supplies remain largely in the hands of the United States Navy.

China is taking some limited steps to set up a blue-water operating capability. In April 2005, Chinese Prime Minister Wen Jiabao opened the Pakistani port of Gwardar which sits at the entrance of the Strait of Hormuz and will host Chinese submarines as part of Beijing's 'string of pearls' strategy to secure naval or intelligence access in regional states (Chellaney, 2005).[4] Speculation abounds that Gwardar will be converted into a transit terminal for Iranian and African crude oil exports finding their way to China. But such ports and facilities are highly vulnerable to highly mobile, offshore precision strike assets such as standoff cruise missiles, extended-range unmanned combat vehicles, deep strike brigades or even sophisticated terrorist and insurgency strikes.[5] Given these scenarios, China would be pursuing a high-risk strategy if it continues to invest heavily in such highly vulnerable maritime strategic assets.

A secondary, but important, factor relating to China's present naval/energy security weakness is that Beijing would presently have difficulties holding its own in any maritime confrontation with other regional powers such as India or Japan in future resource 'wars'. The PLA Navy is ranked as the world's sixth most powerful (after the United States, Russia, the UK, France and Japan) but is relatively obsolete compared to regional counterparts (Nolt, 2002: 325).[6] By 2010, India will have three fully operating aircraft carriers; despite recent rumours to the contrary, China will not yet have any operational equivalent. Japan's Maritime Self-Defence Force is now regarded by many observers to be the region's most powerful navy, apart from the US Seventh Fleet, with modern destroyers, formidable naval air defence capabilities and increasingly impressive surveillance assets that can track and, if need be, could neutralise Chinese submarines conducting incursions into Japan's economic enterprise zones.[7]

Safeguarding against energy supply disruptions

With the reality that US and allied naval power will remain the primary guardian protecting key energy SLOCs and crucial Middle Eastern oil supplies over the foreseeable future, China and other Asian states are weighing alternative strategies to hedge against interruptions in national energy supplies. These include the establishment of strategic petroleum reserves (SPRs), investing in transnational oil and gas pipelines, and diversifying fossil fuel supply sources.

The SPR issue is surprisingly intense in China, given its overall energy vulnerability problems. Opponents of establishing such a reserve argue that China cannot afford to build a SPR large enough to be effective relative to China's vast population and rapidly growing economy. They also assert (perhaps not anticipating China's exploding transport needs) that China's heavy reliance on coal negates the need to implement expensive measures for stockpiling oil or that it encourages excessive oil dependence relative to other energy alternatives (Downs, 2004: 33–34). Nevertheless, the Chinese Government's tenth five-year plan has directed that an SPR build-up in China be initiated by the end of 2005. Japan started its SPR in 1978 and has established an 'obligatory inventory' of 70 days and a de facto stockpile of approximately 166 days. South Korea commenced strategic oil stockpiling in 1980 and had built up 107 days of emergency oil stocks by 2003 (Shin, 2005). The Asia Cooperation Dialogue Energy Working Group, which includes most other small oil-consuming Asian states, has targeted a 30-day emergency reserve for most participating countries. The extent to which this is met, however, will largely depend on the degree of co-ordination that is achieved in ongoing collective regional efforts to build such a reserve.

Constructing new transnational oil and gas pipelines is a second way for Asia–Pacific states to maintain the flow of energy supplies but this is not necessarily more secure than transporting them by sea and may be less profitable. The Central Asia Pipeline project, for example, envisions piping natural gas originating in Turkmenistan through Afghanistan to Pakistan with a possible extension to India. It would be established in conjunction with oil pipelines beginning in Kazakhstan and Turkmenistan through Afghanistan to Pakistan and India. Additional gas pipelines are also projected to be built both between Kazakhstan and China, extending to Japan and Korea, and Turkmenistan and China.[8]

Supply interdiction risks

In all these cases, however, the security risks and commercial costs involved in pipeline construction and maintenance are substantial. These include the relative availability of sufficient oil and gas deposits to justify multi-billion dollar infrastructure costs for pipeline construction and maintenance, prospects for terrorist attacks against these facilities (with recent attacks on pipelines in Iraq and Saudi

Arabia serving as precedents), and the replacement of current regimes in Islamic oil-producing states with more radical governments less prone to dealing with Western and Asia–Pacific markets cancelling contracts or nationalising company assets. For both China and Japan, Iran remains a lucrative, long-term energy collaborator. Iran's own strategic behaviour, however, posits a comparatively high risk of US military interdiction of pipeline routes or the blockade of commercial, energy-related maritime traffic. Such a blockade would present North-East Asian oil consumers with very hard choices in regard to their security relations with Washington.

Two other factors influencing Asian perceptions of energy supply reliability should be noted. Both of these entail disruptions or interdictions to energy supplies that may be largely beyond the control of Asian security elites to contest. First, major Asian governments have only just begun to trust the international energy marketplace and the private ventures that shape it to provide the energy resources needed for national security. Although Japan, South Korea and selected ASEAN states have been involved in energy market liberalisation for some time, China and India, as the two giant regional energy consumers, have just begun shifting their own energy infrastructures from government to private sectors. A study prepared for the Asia–Pacific Center for Security Studies recently encapsulated this problem:

> Although there was general consensus about capital-poor states like China, India, and some ASEAN states adopting more market-based solutions for their energy needs, skepticism persists regarding the possibility of a rapid, regionwide trend toward liberalization of energy sectors … In essence, those economies that are most resource-dependent on foreign sources are more inclined to be reluctant about market-based solutions, yet these are the states that would benefit most from liberalization and deregulation.
>
> (APCSS, 1999)

Moreover, various 'wild card' events could seriously disrupt supplies for weeks or months before Asian governments were able to coalesce with the United States or other OECD states to modify their effects. These may include (but not be limited to) a steep and sudden decline of oil or gas production capacity triggered by terrorist attacks on production capacities in the Middle East, Central Asia or in other energy production centres; the rapid ascendancy of corruption in the extra-regional distribution of oil income (possibly implicating Asian states' own commercial enterprises in the process); or a global economic recession wreaking havoc on energy-producing countries by slicing revenues and creating domestic political instability (Umbach, 2004: 144–145). These contingencies suggest that while globalization does blur the lines of division between domestic stability, national security and international political economics, it is the geopolitical ramifications of events that are often beyond the control of its key players that could shape and traumatize the Asia–Pacific and its energy systems. 'Energy nationalism' is best checked by the identification and application of effective crisis management instruments (such as stockpiling regimes or regional energy consortia arrangements that promote inter-state energy co-operation) that underwrite regional and international resilience during supply disruptions. When such instruments are in place, repeat or more serious episodes like the panic buying of oil by China, India and other Asian states prior to the outset of the 2003 Iraq War are less likely to occur (Herberg, 2004: 371).

Energy source diversification: strategic complications?

Ongoing events in North Korea, Pakistan and other currently or recently designated 'rogue states' highlight the problem of separating national energy diversification strategies from trends in nuclear proliferation, arms sales, environmental security and other developments that may threaten international security and arms control. The political prestige that Asian governments often affiliate with initiating nuclear power programs can lead to powerful incentives to disdain non-nuclear energy solutions (Kessler, 2004). Asian energy involvement with the weapons programs of such 'rogue states' as the Sudan and Burma effectively circumvent US and Western sanctions directed against these countries. Arms control can be undercut by exporting nuclear and missile technology to oil exporters or other energy collaborators. China's sale of DF-3 missiles to Saudi Arabia in 1987 and its subsequent sales of nuclear reactors and M-11 missiles to Pakistan are illustrative; so too is the United States' willingness to lift previous sanctions directed toward India and to provide it with nuclear fuel in return for Indian assurances that it will behave as a 'responsible' nuclear power by 'separating' military from civilian nuclear fuel in its own program. China's and India's relentless consumption of coal and increasing transportation grids have given them the dubious honour of being among the most polluted countries in the world and major contributors to global warming.

Growing fluctuations in the energy diversification strategies of China and India have created largely anarchical and increasingly volatile Sino–Indian relations. China's energy policy toward India is a dichotomous posture: to compete with the Indians for control over key energy supplies while remaining open to co-operative arrangements in other sectors where it deems Indian involvement to be advantageous to holding down its own costs. In October 2004, for example, China outbid India for oil exploration rights in Angola and it has successfully blocked Indian overtures to penetrate Burma's energy sector. However, while China holds a 50 per cent share in Iran's Yadavaran oil field, India holds 20 per cent. Sino-Indian collaboration is also ongoing in the Sudan. China is also sympathetic to India's proposal to create a 'pan–Asian gas grid' – to what extent joint energy collaboration can overcome their legacy of geopolitical competition remains unclear (Hyder, 2005).

Japan's recent energy diversification legacy has not been problem-free but is among the most impressive in the industrialised world. It had reduced petroleum as it primary energy source from 70 per cent in 1970 to just over 50 per cent in 2001 and now uses nuclear energy to generate about one-third of Japan's electricity

(Chanlett-Avery, 2005: 4; see also Kelly, 2005: 278–327. Natural gas consumption provides 13.5 per cent of its energy, with supplies imported largely from South-East Asia but projected to come increasingly from Sakhalin when a joint development project commences with Russia. Its per capita energy consumption is one of the world's lowest for developing nations (using about half the British thermal units, Btu's, per capita than the United States). It has committed substantial funding to develop solar, hydro and other forms of carbon-free, renewable, energy sources (Chanlett-Avery, 2005: 4–5).

South Korea's energy diversity is similar to Japan's but less efficient and more uncertain due to the North Korean factor. Oil provides 55 per cent of its total energy supply, coal 21 per cent and natural gas 10 per cent. The Middle East provides a far greater percentage of LNG exports to South Korea than to Japan. South Korea's nuclear industry is substantial (it has 19 nuclear reactors) but not on the same scale as its Japanese counterpart. Those who support greater engagement with North Korea view favourably the construction of gas pipelines through the DPRK to access Russian and Central Asian energy sources (Calder, 2004). However, the failure of the Korean Peninsula Development Organisation, and the pending status of the Six Party Talks to resolve the North Korean nuclear problem, make such collaboration doubtful. As Kent Calder has observed, 'In the absence of a verifiable nuclear non-proliferation agreement with the DPRK, it is obviously premature to move toward agreement on a trans-North Korea pipeline, from any of the ... prospective sources of Russian gas, even though it would be cheaper than alternatives ... ' (Calder, 2004: 15–16).

Diversification of energy supplies will continue to pose a major strategic challenge for Asian importers. Because the Persian Gulf holds two-thirds of the world's oil reserves to meet global demand over the next quarter century, and may supply up to 90 per cent of the Asia–Pacific region's oil supply by 2010, the dangers of ignoring or not adjusting to the domestic political, economic and social dimensions of energy geopolitics in that part of the world are particularly acute.[9] Yet these dimensions cannot be separated from the international system's overarching and rapidly changing strategic landscape. That landscape must ultimately shape the will and capabilities of Asia–Pacific states to create alternative energy infrastructures to avoid future 'oil shocks' and related energy shortages. Continuing or intensifying dependence by Asia–Pacific states on primarily Middle Eastern oil supplies cannot be justified as a long-term strategy or one that adequately hedges against strategic developments there.

Greater institutional collaboration in confronting energy issues would be a major breakthrough. In this regard, the Association of Southeast Asian Nations' (ASEAN's) Petroleum Security Agreement, requiring ASEAN member states to provide crude oil and/or petroleum products for countries in short supply can be viewed as an instructive precedent (Guoxing cited in Giragosian, 2004). So too would greater joint investment in emerging production locales including West Africa (Nigeria and Angola), Sakhalin and Central Asia, applying formulas designed to minimize the nationalist competitive aspects of such investment. Most importantly, any such collaboration needs to be co-ordinated closely with the world's other great oil consumer, the United States. Excessive dependence on Middle East oil by both the US and Asia–Pacific countries, and the absence of forward-looking mutual strategies forged in appropriate institutions or forums for dealing with this prospect, is a dual recipe for geopolitical disaster. The co-ordinated development of markets and pooling of investments toward long-term, well-considered strategic resource diversification is urgently required to meet the prospect of looming oil shortages (Salameh, 2003: 1085–1091).

Dealing with contingent and structural risks

In the absence of long-term collective strategic planning, Asia–Pacific states will incur a number of contingent and structural risks to their energy supplies that could significantly worsen their strategic positions. *Contingent risks* are unpredictable events that could directly threaten energy supply security: political and military upheavals and accidents at energy production facilities or along energy transportation routes. *Structural risks* include producers' embargos imposed for political or economic reasons, the strengthening of producers' marketing control, environmental ramifications of excessive fossil fuel consumption and lack of energy infrastructural development (Koyama, 2001: 3). The demarcation of contingent and structural risks, however, draws attention to the need for responding to both unanticipated and predictable threats with a sufficiently coherent energy strategy that can realize long-term energy security while reacting quickly and effectively to short-term surprise.

Contingent risks

Political and military risks constitute the most obvious energy security contingency scenarios. Risk assessments must be applied in assigning national security resources and policy to those contingencies deemed to be most likely to threaten supplies. Obvious areas of policy focus include the logistical disruption of oil supplies by terrorism at production sites or along fuel transportation infrastructures and chokepoints; civil unrest or war in producing states; inter-state wars with ramifications for fuel production; explosions or meltdown at production sites; or embargos by other consumer states against designated supplier countries.

Approximately two-thirds of all petroleum produced is shipped by maritime transportation; the remainder is transported by pipelines or by trains and trucks over shorter distances (Rodrigue, 2001: 9). Most of the maritime trade between Europe and the Asia–Pacific passes through the Malacca Strait chokepoint, including one-half of the world's oil and two-thirds of its LNG. Yet its physical features would appear to preclude such traffic. Its egress points are shallow and narrow, impeding navigation as increasingly large ships are employed for oil transport and inviting easy interdiction by hostile elements. Co-ordinated naval patrols have been initiated by Indonesia, Malaysia and Singapore (Operation MALSINDO) in lieu of an offer in mid 2004 by the US Navy to lead a Regional Maritime Security Initiative and another offer by Japan in March 2005 to contribute Japanese Coast Guard patrol boats for strait surveillance and patrolling. Both offers were rejected by the Indonesians and Malays on nationalist grounds (IAGS, 2004; IISS, 2004; Percival, 2005). The effectiveness of current MALSINDO patrolling arrangements, however, is questionable as patrols from one participant state are generally not allowed to cross another state's sovereign waters. Moreover, terrorist groups such as the Philippines' *Abu Sayyef* and Indonesia's *Jemaah Islamiyah* are reportedly showing increased interest in attacking commercial shipping as an easy 'soft target'. In March 2003, ten armed men hijacked the Indonesian chemical tanker, *Dewi Madrim*:

> ... apparently for the purpose of learning to steer it. After operating the ship for an hour through the strait, which narrows to 30 miles in some places, they left with equipment and technical documents. This might be the Maritime equivalent to the Florida flight school where the September 11 terrorists took their flying lessons. (IAGS, 2004)

Combined with the South China Sea (a body of water where unexploited oil and natural gas deposits are thought to be substantial), the Malacca Strait and its strategic passages represent one of the world's most attractive interdiction targets for those intent on disrupting global markets and security.

The Asia–Pacific region is now the world leader in the growth of nuclear energy technologies. There are 56 nuclear reactors or accelerators operating in 15 Asian countries to generate electricity and to meet both medical and development needs (ElBaradei, 2004). Standards of reactor protection, however, vary from country to country in Asia, with the eight reactors in South-East Asia thought to be more vulnerable to attacks or theft of nuclear materials (Ogilvie-White, 2005: 7). While attacks on a nuclear power plant could not lead to a nuclear explosion, the initiation of meltdowns, major fires or conventional explosions could lead to widespread radiation. Attacks on vehicles or trains transporting nuclear materials must also be factored into any security equation. Missing or stolen nuclear materials could be used to construct so-called 'dirty bombs' or radiation emission devices that would contaminate large sectors of urban areas surrounding any reactor or facility. The effects of an aircraft slamming into a nuclear reactor and penetrating its containment walls are unknown but, following September 11, cannot be ruled out as a possible scenario.

All of these nuclear contingent threats have been anticipated by Asia–Pacific states, and mechanisms to deal with them have begun to materialize. Working with the International Atomic Energy Agency (IAEA), the region has commenced the painstaking tasks of accounting for used nuclear materials, retrieving disused sources and shipping them safely back to countries of origin, and training national authorities and technicians to recover radioactive sources in the event nuclear accidents or terrorism does take place. Australia, Japan and South Korea have been prominent contributors to the IAEA's Nuclear Security Fund and China has pledged future financial support. Australia has established a regional project on the Security of Radioactive Sources, and the IAEA will contribute to this initiative with funds earmarked by the US Department of Energy for assistance in establishing 'Regional Radiological Security Partnerships'. Australia, the IAEA, and the USA, together with other Asia–Pacific states – Indonesia, Malaysia, Thailand and Vietnam – are now developing a project work plan (ElBaradei, 2004). The ultimate success of such ventures, however, will depend on how well project policy managers remove traditional barriers to regional co-operation against contingency risks – sovereign prerogatives, corruption and shoddy marketing practices related to infrastructure building and maintenance.

Structural risks

While still significant, structural risks may be less threatening than contingent risks. Illustratively, the embargo politics of the Arab states and OPEC that were applied in the 1970s proved to be counterproductive. At present, the Middle East states and an increasing number of other energy-producing states in Central Asia and Africa are energy-revenue dependent, under increased pressure to re-invest in their energy sectors by growing populations with greater financial expectations, and dependent on Western and Asia–Pacific technologies to diversify their economic bases over the long-term. The energy diversification outlined in the previous sub-section has increased oil/gas-producer states' awareness of their customers' flexibility in reallocating energy supplies. Most producer states would be hesitant to behave in ways that would jeopardize stable revenue flows.

Notwithstanding such greater willingness of energy producers to supply users, structural risks are still capable of generating substantial disruptions in energy flows. If Asian GDP growth continues to increase at current rates (in 1993 Asia constituted 23 per cent of the world's total GDP; in 2010 that figure will have increased to 36 per cent), competition among regional energy importers will inevitably rise. Leaving market forces or unrestrained nationalism to determine the need for oil imports as opposed to alternative oil dependency by energy diversification risks aggravating environmental degradation and defaulting too much of the energy sector's public good to inherently narrow interests. In the era of globalization, it is becoming clear that maintaining critical services in the health, power and telecommunications areas are dependent on governmental co-operation, regulation and accountability. In the energy security arena, '... numerous uncertainties ... require governments to play a role in promoting energy projects which ensure reliable supplies, competitive pricing, and reduced pollution. Government leadership will be indispensable, particularly in attracting and facilitating private sector participation' (Ivanov & Hamada, 2002).

One other structural factor could seriously undermine Asia–Pacific energy security. If an 'oil shock' leads to greater volatility in financial markets, energy-importing states' over-reliance on external oil supplies will be exposed and their overall wealth will come under greater risk. A sustained global surge in oil prices, for example, could lead to a replay of the 1997–1998 Asian financial crisis and to prospects for political unrest and geopolitical rivalries. Over the long-term, effective energy source diversification may mitigate this effect. However, the

Asia–Pacific's current high level of reliance on Persian Gulf crude, and its still underdeveloped project infrastructure network, presently renders the area vulnerable as major net-importing countries China, India, Singapore, South Korea, and Taiwan are most at risk vis-à-vis market forces (Giragosian, 2004).

Conclusion

Energy reliability is a requisite for future Asia–Pacific geopolitical stability. Visible progress has recently been made in regional energy co-operation through institutional action and limited joint collaboration in investing and developing in extra-regional energy sources. Such progress, however, could be overwhelmed by unexpected contingent risks and threats over the near-term or compromised in the more distant future by the region's collective failure to address remaining structural problems in energy supply and protection. Contingent threats equate directly to geopolitics because the power and assets needed to respond to terrorism, resource conflicts and natural disasters with resource ramifications link directly to states' respective national security capabilities. Controlling 'energy nationalism' will have much to do with how much strategic stability the Asia–Pacific region experiences over the next two to three decades. Such control presumes, and cannot exist in the absence of, concurrent implementation of the confidence-building and arms control measures required to dilute incentives for greater militarization in the region and the co-ordination of systematic approaches to dealing with terrorism and other non-state threats.

Resource diversification, infrastructure investment and energy conservation are longer term, more 'structural' strategies but are no less relevant to Asia–Pacific geopolitical stability. Energy security efforts conducted independently by individual Asian economies could be counter-productive to collective energy security management. China, Japan and India, as the region's largest energy consumers, can take the lead in working together to forge a region-wide energy security agenda through regional institutions and in conjunction with international bodies such as the International Energy Agency and the International Atomic Energy Agency. In so doing, they can establish sound energy policy that can be applied to meeting common energy challenges without geopolitical turbulence.

Notes

1 Background is provided by Herberg (2004: 357–361).
2 For an opposing and more optimistic opinion, see Jaffe (2001b), Senior Energy Advisor, James A. Baker III Institute for Public Policy, at http://www.erina.or.jp/En/Research/Energy/Jaffe42.pdf. Jaffe contends that 'it is by no means a foregone conclusion that the nineteenth century pattern of neo-mercantilist competition for territory and diminishing oil reserves need fit analogously with 21st.
3 The 'free-riding' argument was posited by Feigenbaum (1999: 79–80).
4 Other 'pearls' in China's sea-lane strategy include the building of a container port facility at Chittagong, Bangladesh; cultivating close ties with Burma's military junta in order to gain more access to that country's ports located along the Indian Ocean; and exploring with Thailand the long-term construction of a canal (costing an esti-

mated $US20 billion) across the Kra Isthmus that would allow Chinese commercial vessels to bypass the Strait of Malacca. Defence consultant Booze Allen Hamilton recently prepared a classified report for the Pentagon outlining this strategy in some detail see Gwertz (2005). For an assessment of other regional security actors' basing activities in the Indian Ocean, consult Berlin (2004: 239–255).

5 Relevant scenarios are outlined by Krepinevich (2002).

6 Citing various public intelligence estimates.

7 For a quantitative comparison of Japanese–Chinese naval capabilities, see Lind (2004: 100). For a useful general assessment of Japanese force modernisation trends, see Dupont (2004: 25–36).

8 For a comprehensive itinerary of proposed pipeline projects from Central Asian states to other regional centres, see Finon *et al.* (2000: 12–13).

9 This oil consumption estimate is provided by Ji Guoxing of the Shanghai Institute of International Strategic Studies and cited by Giragosian (2004).

10 Economic dimensions of energy security in the Asia–Pacific

Stuart Harris and Barry Naughten

Energy, and in particular, energy security, has re-emerged high on the policy agenda of countries in Asia–Pacific as elsewhere. Interestingly, greater attention in Asia seems to be on energy security rather than on the macroeconomic effects of short-term price changes; in the 1970s and 1980s, for most countries, the effects of the high oil prices were more salient – Japan perhaps excepted – and more concern was expressed about the effect of high and volatile oil prices on economic activity, especially recessions and excessive inflation.

This may be a result of the global intensity of energy use having fallen. Or perhaps that the earlier impact was less than thought and the link with the stagflation then prevalent was exaggerated; or the current price levels in real terms are still less than the peaks of the 1970s and the 1980s; or that current low interest and inflation rates enable the global economy to absorb high energy prices more easily. There is some truth in each of these explanations but oil remains a critically important strategic commodity for world economic activity and especially for the Asia–Pacific countries.

In addressing energy security in the Asia–Pacific region, we deal mostly with the oil market. Although the gas market is growing rapidly in importance and, as we discuss later, has security characteristics that differ from oil, its price has tended to follow the oil price, often explicitly in long-term gas contracts. We ask why the current market situation has emerged? What are the prospects for the fossil fuel energy market? In particular, what does it all mean for the Asia–Pacific region? And, specifically, what do we mean by energy security and how can the Asia–Pacific countries increase their energy security?

Why has the current oil market situation emerged?

The Asia–Pacific oil markets are part of the global market and we look first at the global market situation for oil before considering the specific Asia–Pacific dimension. The gas market is still largely regional but will become less so as its use increases. Prices of oil in particular on the world market are currently both high and volatile. At their recent low point in 1998, oil fell to around US$10 a barrel but since 2001 oil prices have generally trended upwards. Prices from March 2004 have generally been over 80 per cent above prices 2 years earlier – in real terms (see Figure 10.1).

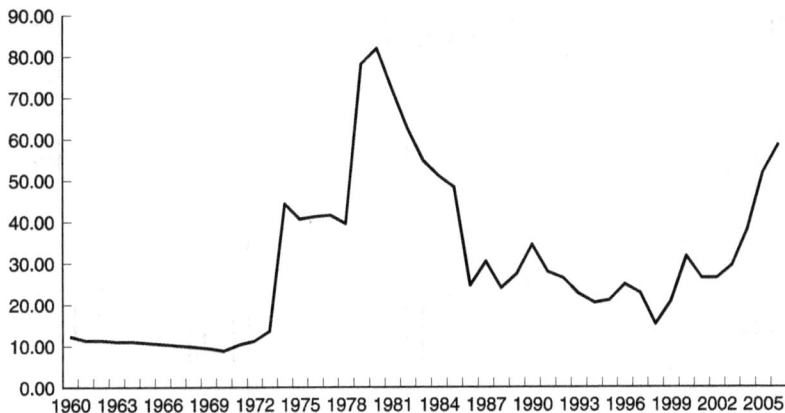

Figure 10.1 Crude oil price, Brent, 1960–2006 (first quarter only), 2004 US$/b.
Sources: BP Amoco, USDOEEIA (updated by the authors).

The major increase in pressure on oil prices in 2003 and onwards reflected a mix of demand, supply and speculative factors. Important in recent years has been the high demand for oil linked to the robust US economy and to the surge in China's economic growth.

Oil production, processing and transport capacity is based on investments the time scales for which are lengthy and the financial magnitudes large. Although investments are made on judgements about future profitability, oil prices have varied greatly over the last 30 years with a difficulty to determine trend – varying, according to the International Monetary Fund (IMF), from US$8 to US$96 a barrel in 2003 (IMF, 2005b: 159). Following the 1970s and early 1980s price rises investment in the industry was heavy and production capacity, especially in non-OPEC countries, expanded rapidly. The response to what investors saw as a period of over-investment and excess capacity has been 20 or so years of under-investment (IMF, 2005c: 37). Investment in the key low-cost OPEC regions, from which the international oil companies (IOCs) have since the early 1970s been largely excluded, tended to languish. The sensitivity of exploration expenditure to the oil price is shown in Figure 10.2.

Meanwhile, over-investment during the early 1980s occurred in other sectors such as pipelines, shipping facilities, stocks and particularly in refineries. With low profitability in the refining sector until demand surged in 2003, little refinery capacity was added in the later 1990s. The current situation has reflected particular shortages of US and to a degree Asian refining capacity suitable for processing the heavier, high sulphur, oils that Saudi Arabia could add to market supply to meet the demand growth (al-Yabhouni, 2005).

Consequently, when global economic activity surged unexpectedly, as in 2003 and particularly in 2004–2005, the industry soon came close to capacity limits. Given the lack of spare supply capacity, especially in Saudi Arabia, the key

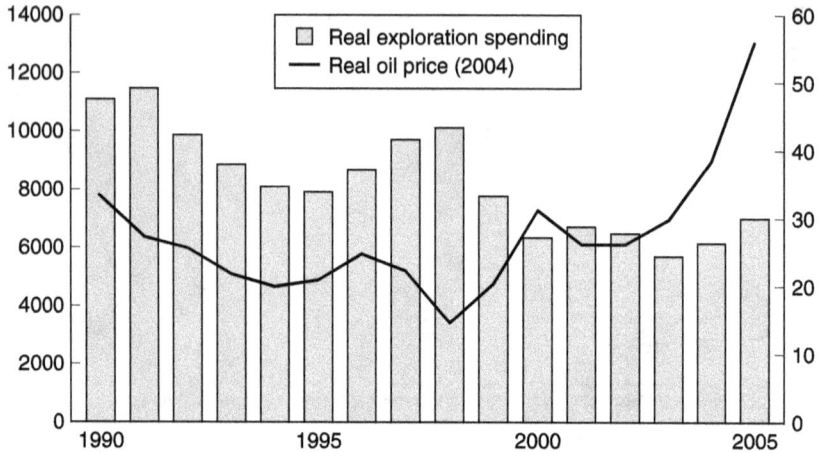

Figure 10.2 Exploration spending among major oil and gas companies.
Source: Osmundsen, P., Mohn, K., Misund, B. and Asche, 2006.
(*Data source*: Deutsche Bank: Major Oils 2004, deflated by US CPI).

'swing producer', geopolitical uncertainties in some oil supply areas – Venezuela, Nigeria, Iraq and Russia – helped sustain the speculative pressure. This led not only to higher prices but to considerable volatility in oil prices.

In summary, important in the current market situation has been a global economic boom leading to surging demand, an oil market lacking in capacity to meet that demand, and uncertainty in the energy market generally.

Market supply and demand

Moreover, price increases and price volatility have revitalized concerns that in circumstances of capacity limits, importers are vulnerable to political or other arbitrary interference in the market. Added to these anxieties has been speculation that the higher oil prices are indicative of long-term physical supply limitations. The International Energy Agency (IEA) has addressed that question in a different form that has also become a focus in the energy debate; that relates to when oil production will peak. There have been various estimates, including a number that the peak would be reached in the next few years. The notion of a production peak is preferable to loose references to 'when will the oil run out' but the presumption that it is close to realization leads to a belief that the international 'scramble' for supply sources in a declining level of oil production will intensify greatly and that, only at best, will oil prices rise sufficiently to ration what is being produced. This debate is not new: it was current in the late 1970s, and at the global level production did reach a temporary peak in 1979; not until 1994 was the production level of 1979 reached again. This shows that a global peak in oil production can be as much a demand and price phenomenon as a supply phenomenon. As we discuss later, the basic conclusion of the main institutional analysts would seem to be that, while economically recoverable fossil fuel resources are finite and will eventually constrain production, this is not likely in the near future.

For example, despite expecting oil demand to be some 60 per cent higher in 2030 (the end point of its projection target period) than now, the IEA does not expect oil production to peak in the next quarter century at least. OPEC countries, mainly in the Middle East, are expected to meet most of the added demand.[1]

The expected annual growth globally in energy demand is lower than the average of the past three decades (1.7 per cent compared with 2 per cent) and oil demand growth is projected to be slightly less still (1.3 per cent). Whether this demand growth is met without major oil market problems will depend, however, upon very large investments in oil fields, tankers, pipelines and refineries. This proviso is important; as Robert Mabro (1998: 4) notes, whenever prices rise, one needs to look at capacity for an explanation, and ask why capacity is insufficient.

While the level of oil prices is a major influence on investment decisions, the volatility of prices is also important since decisions on investment are often deferred in a volatile market pending greater clarity about where the market is going. So the future volatility of oil prices as well as their level will affect production and the level of investment in exploration and development and in transport and refinery capacity.

The magnitude of the investment required to meet the growing energy demand to 2030 is very large. The IEA has estimated the total investments required in the oil and gas sectors of the energy industry over that period at US$3 trillion (IEA, 2003b: 25). At the global level, this is not an unduly difficult prospect to envisage, given the size of global savings. It does depend, however, upon the political and economic conditions that exist. This is now more important than in the past since, in the IEA's view, more investment funding will need to come from private, commonly foreign, funding, while at the same time, there will be greater dependence upon oil from areas subject to national government control and restriction.

Global economic impacts

Before looking at the effect of high oil prices on the Asia–Pacific countries, we again look first at the global picture. High oil prices will have a downward effect on global economic activity; but the question is: how much? For net oil exporters, adverse direct and indirect domestic economic effects will be offset by income transfers from increased export earnings. For oil importers, there is no such offset for direct and indirect effects of high oil prices.

In practice oil prices in real terms in 2004–2005 were still well below the levels of the late 1970s and early 1980s. Moreover, since oil prices are traditionally defined in US dollar terms, to the extent that the US dollar has depreciated, the domestic effects are reduced.

The link between oil prices and macroeconomic effects is a complex interrelationship and the impacts follow various channels. The IMF notes that in the past a permanent $5 a barrel increase in oil prices would lower global GDP growth by up to 0.3 per cent; in practice, the impact in 2004 was less than feared – due mainly to strong global growth and flexible economic policies (IMF, 2005b). The IEA has since estimated that a 10 per cent rise in the oil price will reduce GDP growth in OECD countries by 0.4 per cent in the first 2 years[2] with consequent adverse effects on unemployment and inflation (IEA, 2004a).

The reduced energy intensity in economic activity and the smaller share of oil in total energy supply have helped to limit those effects. We noted earlier that oil price increases were commonly linked with recessions, inflation and slow economic growth, as with the high oil prices in the late 1970s and 1980s. While some increased impact is probable if prices remain high, evidence from past oil price increases is that the contribution of price changes themselves to recessionary, inflationary and economic growth impacts was not large (Barsky & Killian, 2004: 129).

Consistent with the IEA, the IMF estimates that in 2005–2006 global gross domestic product (GDP) would slow by 0.7–0.8 percentage points and possibly higher, relative to 2004, with oil prices a contributing factor.[3] In practice, how important oil price increases are depends to a large extent upon external conditions: whether the global economy is booming or in recession; whether expectations reflect confidence or pessimism; and the scope for policy flexibility such as provided at present by low interest rates and low inflation (IMF, 2005b). Such conditions, of course, are not readily available to less advanced developing countries with high debts.

Impacts on the Asia–Pacific

The Asian Development Bank (ADB) (2005: 3) noted in September 2005 that 'resilience to higher oil prices has been a notable feature of developing Asia over the past few years'. Nevertheless, the macroeconomic impacts on the Asia–Pacific affect different countries differently. ADB data indicate that the region, as the ADB defines it, imports 44 per cent of its oil.[4] Members of the region export some 11 per cent of the world's oil and import 21 per cent of the world's oil. Net exporters – mainly among the Central Asian states – have gained in income terms from the high oil prices whereas net importers will have lost income; these impacts will show up in upward or downward pressures on their respective exchange rates and/or their balance of payments results.

While we noted earlier that the global impact of oil prices has not so far been great – this is not a precise guide to the impact on the Asia–Pacific region where oil is a larger element in its overall economic activity. The ADB does note that for many oil-importing countries, the effects may start to become serious as import bills increase, and with some signs of inflation moving upwards in some countries and budgetary costs of energy subsidies in a number of countries becoming large.

Asia's heavier reliance on oil in its overall energy supply and, to a degree, to lower energy efficiencies and to more energy-intensive patterns of economic activities, means that compared with oil consumption representing 1.6 per cent of

GDP in the industrialized countries, in the Asia–Pacific region it is equivalent, on an exchange rate basis of GDP measurement, to 4.5 per cent of GDP (ADB, 2005). This is still considerably less in Asia than the 7.3 per cent 25 years earlier. Moreover, on a purchasing power parity basis, it would be significantly lower still, differing much less from the industrialized countries (Maddison, 1998: 155; Harris, 2004: 63).

Nevertheless, despite the high oil prices, economic prospects for the Asia–Pacific region have been assessed as generally favourable with inflation rates mostly remaining subdued. The World Bank foresees East Asia and the Pacific continuing to grow at the high rates experienced since the end of the 1997–1998 financial crisis (World Bank, 2005: Appendix, Table A1). So too does the ADB (2005). A factor maintaining high oil prices but also economic growth in Asia is expected to be the sustained regional economic growth linked to growth in China (ADB, 2005). While US macroeconomic imbalances continue, the US will also contribute to growth and to high oil prices. The US imbalances will also pose longer term risks for the global and regional economies, with increasing US oil imports putting downward pressure on its real exchange rate and upward pressure on interest rates. Given, however, the lower average levels of energy efficiency in the Asia–Pacific, there may be some adverse effect on the relative competitiveness of Asian countries in competition with the less oil-dependent industrial countries.

While factors such as oil-import dependency, efficiency in use and the structure of GDP critically shape how high oil prices affect these economies, the ultimate oil price impact will also depend upon the extent of the price increases and their duration. The ADB has analysed the effect of oil price increases on Asia–Pacific economies according to two different assumptions about extent and duration. The price increase assumptions range from $10 to $20 and the duration from 1 year to 1.75 years. On these assumptions, the analysis shows that the

Table 10.1 World oil supply (Mb/d)

	2004	2010	2020	2030	2004–2030*
Non-OPEC	46.7	51.4	49.4	46.1	0.0
OPEC	32.3	36.9	47.4	57.2	2.2
OPEC Middle East	22.8	26.6	35.3	44.0	2.6
Non-conventional oil	2.2	3.1	6.5	10.2	6.1
World†	82.1	92.5	104.9	115.4	1.3
Shares (%)					
Non-OPEC	56.9	55.6	47.1	39.9	–
OPEC	39.3	39.9	45.2	49.6	–
OPEC Middle East	27.8	28.8	33.7	38.1	–
Non-conventional oil	2.7	3.4	6.2	8.8	–

IEA (2005). Reference case projection IEA (2005: 90).
*Average annual growth (%).
†Total includes 'processing gains'.

downwards impact on GDP growth for the region as a whole will range between 0.6 and 1.1 per cent. Obviously there are major differences across countries; the most adversely affected are the Philippines, Singapore and Thailand, the GDP impact on the latter ranging from 1.7 to 3.3 per cent (ADB, 2005). Poorer countries such as Laos and Cambodia would be more severely affected.

Since these estimates were made, prices have continued upwards. Yet for most sub-regions the adverse consequences for industrial economies and for regional growth have been adjudged small at present. With the likelihood that average oil prices will remain higher than in the past for the longer term, however, the consequences for regional growth will be more serious.

How will the energy market evolve?

In a rules-based competitive market, open to producers and consumers on a comparable basis and given peaceful international relations, the issue of physical supply security would be of limited interest. Supply and consumption would be balanced by price – and available supplies would be allocated by price to those willing to pay the market clearing price. Price volatility could still be a problem but even that in a fully competitive market should be mitigated by movements in stocks and use of financial instruments.

To a degree, in its distribution function the oil market does operate as a competitive market – prices before taxes, transport costs etc. differ little globally from country to country. Even if a country produces a lot of its own oil, its price for domestic purposes – unless strictly controlled – will generally follow the world price.

Nevertheless, there are limits to the effectiveness of the oil market in practice. The cyclical nature of economic activity is one, given significant gaps in the availability of accurate and timely supply, demand, trade and stocks data, among other things, with consequences for optimum decision-making on production and ultimately production and infrastructure investment. Even were the international oil companies to make more of their reserves data and the assessment basis information available, estimates suggest that only 14 per cent of proven reserves of oil and gas are fully available to international oil and gas companies (Zanoyan, 2004: 2). For oil proven reserves alone another estimate is that the traditional 'oil majors' now account for little more than 6 per cent of global 'proven reserves' (Renner, 2003). The remainder is under the control of states and their national oil companies where for some – perhaps many – reserve figures are often effectively state secrets.

Output of some non-OPEC oil supply sources, notably in Europe, has declined and this leaves the supply side of the market more dependent upon geographically concentrated supply sources in areas either judged geopolitically uncertain or under national government influence. Oil production increasingly depends upon OPEC, Russian and Central Asian countries; where government controls and restrictions will often limit how far production and investment decisions can respond to prices and price expectations reflecting underlying market conditions.

Moreover, the oil traded on the spot market on any day is proportionately very small and is therefore subject to considerable fluctuation. The establishment of a reference price system was aimed at limiting the market information problem (Mabro, 2005). This price, however, also reflects prices on futures markets that include expectations which may be 'guided' by OPEC statements, while oil futures contracts have become financial instruments as well as product price indicators. Consequently, they are now less satisfactory as indicators of the underlying market situation.

Some consequences of these market imperfections are more evident in some circumstances – as in global booms – and less in others. This is what Skinner and Arnott refer to as context – the state of global markets, of regional stability and above all of the relationship between buyers and sellers of the commodity. Context also relates to changes in attitudes. In the 1980s, OECD countries had banned the use of natural gas (seen as a premium fuel) for electricity generation whereas now it is a preferred fuel for that purpose (EUROGULF, 2005: 25–27).

OPEC countries remain important as global supply sources but, until recently, less so than they did some two or three decades earlier. They currently control some 70 per cent of the world's proven reserves, producing about 40 per cent of total world output (and supplying over 50 per cent of internationally traded oil) (IMF, 2005b: 161).

The role of OPEC countries is likely to increase and to reflect a greater concentration of OPEC production in the Middle East.[5] Although in 2004–2005, OPEC was producing close to its capacity, it has reserves that, given the necessary investment, would enable capacity to be developed. OPEC production costs are still substantially lower than those for non-OPEC production.

High prices reduce demand but also increase investment and output.[6] The supply effect is likely to be more apparent in the higher cost non-OPEC sector. Hence, under the high price assumption, the OPEC share actually declines from its present share and, given the volume decline, so might OPEC revenues. For those, including some linked to the US Administration, with an interest in lower prices and wishing to undermine OPEC (Morse, 2004) this would actually increase OPEC's market share, as is evident from the data in Table 10.2. To avoid this, action would need to be taken to reduce demand through a sustained process of fuel efficiency improvements and lowered global consumption.

One rationale for oil-importing states to engage multilaterally in fuel efficiency or demand management policies is as a counter to any 'market power' wielded by OPEC.[7] That such 'market power' is significant has long been argued by Morris Adelman, a leading US energy analyst, among others (Adelman, 2004: 18). Even if that is disputed (see Mabro, 1998), it is possible to envisage that such market power will emerge as high-cost non-OPEC oil is depleted and the OPEC share of supply increases. Of course, there are also other reasons for policies encouraging energy savings and efficiency, especially with respect to benefits from addressing global and regional environmental externalities.

In assessing reasons for current high oil prices, a blame game has emerged. As noted earlier, for Morris Adelman, and for some other US commentators, the

Table 10.2 World oil supply IEA (2005): scenarios as at 2030 (Mb/d)

Scenario:	2004	2030				
		Reference scenario*	'Deferred Investment'†	'Global Alternative'‡	Difference compared with reference scenario	
		RS	DI	GA	DI	GA
Oil price (2004US$/b)						
Averaged over period		$37	$45	$33	8.0	−4.0
2030		$39	$52	$33	13.0	−6.0
MENA*	29.0	50.5	35.3	44.6	−15.2	−5.9
OECD	20.2	13.5	14.0	ni	0.5	Ni
Rest of the world	30.8	41.2	44.5	ni	3.3	Ni
Non-conventional oil	2.2	10.2	11.6	ni	1.4	Ni
World	82.2	115.4	105.4	103.4	−10.0	−12.0
Shares (%)						
MENA*	35.3	43.8	33.5	43.2		
OECD	24.6	11.7	13.3	ni		
Rest of the world	37.5	35.7	42.2	ni		
Non-conventional oil	2.7	8.8	11.0	ni		

*Middle East and North Africa.

long-term OPEC impact on production and prices is the essence of the oil problem. We also noted that there are counterviews. OPEC blames the shortage on surging demand and under-investment, particularly in refining capacity in consuming countries.[8] The IEA argues, however, that OPEC should put more OPEC oil on the market (Mandil, 2005), although the refinery bottleneck remains.[9]

The assumption that this is possible is reflected much more substantially in the institutional analyses (see Table 10.2 where Middle East and North Africa (MENA) largely consist of Middle East OPEC supply). Since 2003, both the IEA and the US Energy Information Administration (EIA) have revised upwards radically their long-run expectations of the price path for crude oil over the period to 2030 while the IEA's 'Global Alternative scenario' postulates a major effort in demand-side policies to reduce consumption of oil, gas and hydrocarbon energy that implies lower demand and reduced prices.

In the IEA's projected OPEC supply is a residual between estimates of global demand and of non-OPEC supply and is sensitive to assumption variations. OPEC, it postulates, could account for some 50 per cent of world output by 2030, compared with some 39 per cent share in 2004. This equates to an increase from 32 mbd in 2004 to 57 mbd in 2030. Its reserves would be adequate for such an increase but, among other things, there are important technological constraints on how fast oil output can be increased without damaging the resource base, particularly where secondary recovery is underway; hence there must be questions about whether that outcome is probable.

The IEA addresses this issue in its 'Deferred Investment scenario'. As Table 10.2 shows, this would reflect an increased oil price (in 2004 oil prices) of around $50/b by 2030 and a consequential reduction in supply to marginally less than MENA's share in 2004.

Projected OPEC supply is close to the maximum level of OPEC output share in the early 1970s. Yet the evidence suggests that OPEC's capacity to influence the market is stronger in times of boom and weaker when world economic activity is in recession (Skinner & Arnott, 2005a). Moreover, as prices rise internationally, non-OPEC oil sources, including non-conventional resources such as the Canadian tar sands, become more economically viable, reducing OPEC's market share.

The future trend in oil prices is uncertain. US analysts and derivatives traders foresaw the possibility of price spikes of up to up to $105 a barrel in 2005–2006 as did those advising CNOOC in its bid for Unocal.[10] Even with such a spike, however, sustained prices at that level are highly unlikely within a normal investor's time scale. Current prices are well above trend and prices are not expected to rise substantially from current levels in the foreseeable future. Were the longer term prices to approach the recently increased US EIA 'high price' scenario of US$48/barrel, more oil could be drawn from non-OPEC sources with higher marginal costs. If such a price were the norm, the physical supply aspects of the 'energy security' problem would (perhaps paradoxically) be relieved since total oil consumption would be less, as would import dependence for many countries, supply would be more diversified and, in particular, dependence reduced on supply from the Gulf.

There is little consensus on the long-term trend in oil prices. Many analyses expect the oil price to fall, ultimately from high levels of 2005/2006, reflecting new oil coming onto the market, up to perhaps 2010 but probably not back to pre-2003 price, given that marginal production costs are expected to have risen. The IEA, analysing supply and demand balances, assumes a reference case of US$29

Table 10.3 World primary energy demand in alternative policy scenario*

	2003	2030	2003	2030	2003–2030	Difference with reference scenario in 2030 (%)
	Mtoe	Mtoe	Shares (%)	Shares (%)	% pa	
Coal	2,582	2,866	24	20	0.4	−23
Oil	3,785	4,967	35	34	1.0	−10
Gas	2,244	3,528	21	24	1.7	−10
Other	2,112	3,297	20	22	1.7	8
Total	10,723	14,658	100	100	1.2	−10

* Condensed from IEA (2005: 270, table 8.4).

in 2003 by 2030 and at $35 a barrel as a higher longer term price; as noted, the EIA has hedged its bets with a range now from $21 to $48/barrel (IEA, 2005c). One difficulty of predicting future oil price trends is that, in large part, the outcome will depend upon large investments in capacity that may not occur but which are themselves dependent upon investor expectations of future prices. And if past experience is a guide, the bulk of this investment will have to come from publicly traded companies (Zanoyan, 2004: 3).

Large investments are needed simply to maintain output levels – more is needed to provide for output growth. If the argument holds that such investments are now more financially and politically risky, the expected future oil price will need to rise significantly to attract sufficient investment (Pindyck, 1991: 1110–1148; Dixit, 1992: 107–132). Such considerations also emphasize the rationality of 'waiting' for better information about the future before investing heavily in large, lumpy and long lead-time projects such as intercontinental and international oil and gas pipelines, or nuclear power stations and their associated infrastructure.

While oil demand growth is expected to slow, demand for natural gas will grow rapidly, rising faster than other fuels, and is expected to increase its share of global energy consumption from 21 per cent to 25 per cent by 2030. At present, pipeline transport accounts for some 70 per cent of total trade. LNG, however, is expected to increase more rapidly and to account for 50 per cent of trade by 2030. Market flexibility is also increasing as spot trading increases – already accounting for 11 per cent of total trade as against only 2 per cent in the late 1990s. Proved gas reserves are said to be adequate for at least 40 years to meet growing demand. Prices, while likely to follow a similar pattern to oil prices, would then remain competitive in major uses.

Energy security

Given market developments, including expectations of continued rapid growth in Asia–Pacific demand, the resurgence of interest by governments in energy security is not surprising.[11] Yet what is meant by energy security is far from obvious. We noted earlier that, in economic terms, security problems imply what economists refer to as market failures or market imperfections. Moreover, like all forms of security, it is in part a question of perception.

Nevertheless, for present purposes, energy security would involve confidence that the country concerned would have at all times an adequate physical supply of energy at reasonable cost. But what is a reasonable cost is unclear; relative to other importers or markets? To some 'competitive market' ideal? Or to some expected long-term trend?

Measures to improve energy security are like insurance and, like insurance they entail a cost. How much of a cost for insurance purposes is warranted is a question for which no clear answer is possible – if only because of the perception nature of security and expectations that relate to future oil prices, often exacerbated by speculation.

Although there are various phenomena that give rise to a lack of confidence in supply security, in their effect they eventually resolve into two: physical supply interruptions and price instability. These are not completely separate since physical supply interruptions may, and normally do, transform into price effects.

It appears that for those economies dependent upon imported energy, the major fear is that of physical interruptions as a result of political or military interference in access to supply sources. This is a concern that the Chinese have in the light of the US involvement in Iraq as a policy to reshape the Middle East. Moreover, they, like the Japanese in particular, have more general concerns with respect to their dependence on Middle East oil supply sources and sea-lane transport. Following 9/11, wider anxieties have been expressed over possible terrorist attacks on energy facilities and, particularly in the Asia–Pacific, attacks on seaborne or pipeline transport of oil and gas.

Evidence of past geopolitical events in the Middle East suggests they have commonly had a substantial impact on oil prices. More or less sharp or prolonged peaks (1978–1980, 1990–1991, 2004–2005), sharp falls (1986) and deep troughs (1997–1998) in price have occurred from time to time, some attributable at least partly to Middle East 'instability' although not always to major supply interruptions.

OPEC does, of course, attempt to manage the market by varying overall production levels. It has done this in the past by maintaining a degree of unused capacity that provided production flexibility up, as well as down, when necessary to maintain a degree of price stability. It plans to maintain a level of spare capacity in the future for this purpose although there is some doubt that its planned level will be adequate (IMF, 2004).

The possibility of supply interruptions cannot be easily dismissed, if at times it seems exaggerated. Moreover, in oil industry history, there have been many deliberate supply disruptions elsewhere than in the Middle East. Many countries, consumers and producers, have used oil and gas leverage, not always successfully. The US and the Netherlands used it against Japan just before World War II; the US and its western allies used it against the Soviet Union during the Cold War, restricting imports of Soviet oil and gas in the 1960s, and at times against South Africa, Serbia and Haiti; they also at times embargoed oil imports from Libya, Iran, Iraq, Sudan and Burma (Alhajji, 2005). Energy leverage has been used on a number of occasions between Russia and the Baltic and CIS states.[12] China itself cut supplies of oil to North Korea for 3 days in early 2003, ostensibly an accidental cut but seen widely as a message to Pyongyang to co-operate on its nuclear weapons program.

A second fear is increasing price volatility. We noted earlier the apparent decline in concerns about macroeconomic effects of price volatility. This may be in part at least due to reduced energy intensity of economic activity in developed countries together with improved macroeconomic policies and instruments. The macroeconomic consequences of unexpected and rapid jumps in prices can still be adverse for Asian countries, oil exporters as well as oil importers, where policy frameworks are often weaker or more constrained. The problem is also, as we saw earlier, that volatile prices also inhibit investment and that low or delayed investment can contribute to future price volatility. Efforts to reduce price volatility are still important therefore for Asia as a whole.

The question of price volatility has short- and long-term aspects. As we noted, physical supply interruptions commonly emerge as price variations. The causes of physical supply interruptions are likely to be due to political actions or to geopolitical upheavals in supplying regions or to the exercise of market power.

Whatever in the past might have been the exercise of market power by energy suppliers, the oil price rises experienced in 2003–2005 were not due to deliberate physical supply interruptions. We noted earlier concerns about under-investment. Insufficient capacity became evident in the face of a demand boom and longer term market uncertainty, leading to excessive speculative reactions. The role of speculation is exemplified by the May 2004 oil price response to attacks on Saudi Arabian oil installations; these attacks had only limited impact on oil facilities – yet traders' fears led to sharp price increases (Barsky & Kilian, 2004).

Nevertheless, ultimately the current situation is a longer term price effect that, given a demand surge, reflects under-investment, over-investment, or misallocated investment due to misreading the commonly blurred market signals or because government controls limit the ability of investors to invest optimally or reduce their incentives to do so. Moreover, what is not known is how far the industry will invest sufficiently in the future in capacity levels in production, refining and transportation above present levels to provide a surge capacity. If what the industry sees as an equilibrium level of capacity falls short of such a capacity, price volatility may well increase.

When oil prices rise, they commonly give rise to concerns that traditional energy supplies may in practice be approaching exhaustion. This leads to various geopolitical reactions and to something of a scramble to ensure control of available energy sources. In considering the resource exhaustion issue, we start with the question of oil reserves, a matter on which there is considerable debate and dispute In part this is because the argument is commonly about the concept of 'proven reserves', an economic as much as a physical or technical one. It is not helped by different concepts being used – such as 'possible' or 'probable' reserves, and 'ultimately recoverable resources'. And not all measures of 'proven' reserves apply the same criteria.[13]

To 'prove' reserves is costly and so, provided enough reserves have been proved to support oil industry activity – one industry rule of thumb is that proven reserves should be sufficient to meet 15 years' demand – the incentive to prove additional reserves diminishes. Resource economists point to many previous unfounded claims of fossil fuel exhaustion – for both coal and oil that began in the nineteenth century (Jevons, 1906; Adelman, 2004: 18).

Morris Adelman has been quoted as saying that there is no oil exhaustion problem and that oil resources are not dwindling given the availability of improved data and technological improvements. In support he notes that in 1970, non-OPEC countries had about 200 billion barrels of oil remaining in proven reserves. In the next 33 years those countries produced 460 billion barrels and yet in 2003 had proven reserves of 209 billion. For OPEC over the same period, proven reserves rose from 412 billion barrels to 819 billion despite production of 307 billion barrels (Adelman, 2004). A study for the APEC Energy Security Initiative notes that in 1980, the reserves to production ratios for oil and for gas were 29 and 58 years, respectively; at the end of 2003 they were 41 and 67 years, respectively (ABARE, 2005: 9). The International Energy Agency has indicated that there will be no basic resource availability problem over their projection period of the next three decades at least. US Geological Survey (USGS) data are cited by OPEC showing that since 1940 when the USGS estimated the world's ultimately recoverable resources at 0.6 trillion barrels, consecutive rises in its estimates have taken it to 3.3 trillion barrels (OPEC, 2004: 11); this would suggest that there is considerable potential for discovering more reserves of fossil fuel in the future.

Most oil reserve figures do not include non-conventional oil. Given the level of prices now operative and expected, and the technological changes achieved, the exploitation of some non-conventional oil has become economic. The economically recoverable reserves in Canada's tar sands, extra-heavy oil deposits in Venezuela and shale oil in the US, at around 600 billion barrels, are considerable, around half those of conventional reserves. Given, however, the higher costs of production and the substantial environmental implications of their exploitation, how much of those reserves can be recovered cost-effectively and sustainably remains uncertain. Nevertheless, the IEA argues that perhaps 10 per cent of world oil production could come from non-conventional oil by 2030 (IEA, 2004b: 102).

Given the long-time horizon widely accepted by energy analysts until the fossil fuel base may constrain potential supplies, discounted present values of those future oil prices are sufficiently small that for energy security decisions today, exhaustibility is not yet, in commercial terms, a substantial issue. For governments, the issue would be different.

Options for reducing energy insecurity

In considering what can be done to reduce energy insecurity, it is useful to review why insecurity has emerged. For countries in Asia this comes from rapidly rising energy demand, increasing dependence on imports in an uncertain international environment and a volatile and high-priced market, and domestic limitations in many producing countries on energy production, processing and transport.

A variety of general measures would have beneficial effects in limiting both short- and long-term energy insecurity, whether physical interruptions to supply or through price volatility. These include energy conservation, increased storage, improved processing and transport facilities, diversification of supply and energy sources (including sources closer to home), increased domestic production (where possible) including of substitutes, removal of market impediments, better data and information and its increased transparency, good relations with supplying countries, and increasing international co-operation.

Conservation

Energy conservation – encouraging measures to reduce oil and gas consumption, in particular by increasing the efficiency of energy use – is perhaps the lowest cost way to provide energy security insurance for both short- and long-term benefit. The vehicle fuel efficiency and fuel-saving picture for the Asia–Pacific region, whereas in the industrialised countries, oil will increasingly be used primarily for motor transport, is mixed. On the one hand, Japan and Hong Kong have long had high rates of tax on transport fuel. On the other, fuel tax rates in China and South-East Asia have been among the world's lowest but in March 2006, in China these were brought closer to market prices.[14] At the same time, for new vehicles, China announced reinforced fuel efficiency standards already promulgated in 2004 that are tighter than those of the US and are specifically inclusive of SUVs. It has also begun to introduce a nationally uniform rate of tax on transport fuel.

Nevertheless, even without egregious examples of moving in the opposite direction, as with SUVs in the US and elsewhere, few would expect conservation alone to reduce fossil fuel use absolutely in the near future although, as we have seen, this did happen following the 1979 price with the help of active governmental responses. It would help, however, and contribute to reducing global warming, a condition increasingly seen as a security problem in parts of Asia.

As already noted in Table 10.2, in the IEA's 'Global Alternative scenario' consumption of energy is significantly reduced by investment in energy efficient end-use equipment and infrastructure. As indicated in Table 10.3, this scenario implies a reduction of 11 per cent in total energy use as at 2030. The scenario also involves significant fuel switching away from coal towards gas. This is especially the case in China (IEA, 2005: 273).

Stocks

Emergency stocks are seen by the IEA as a major weapon against energy insecurity. In September 2005, IEA members collectively agreed to release 60 million barrels of oil and oil products over a month to offset the price effect of Hurricane Katrina. In the region, Japan and Korea have released stocks from their emergency stockpiles in accordance with their membership obligations in the IEA. Some years ago, although not an IEA member, China also decided to establish an emergency stock mechanism. Filling, which is expected to take some years, began in 2005 (Bi, 2005).

The IEA action is interesting because there has been a reluctance to release stocks to deal with high prices – the US Administration decided not to do this in 2004 even as prices doubled and Japan similarly would not release stocks to offset price rises. Even EU energy officials who saw stock releases as a counter to price spikes were denied the authority to release stocks for price purposes when, in 2002 they sought the authority to do so.

The logic is that to gain the full benefit of stock releases stocks should be released co-operatively with others. If a single country does so unilaterally, the market effect is widely dispersed and all market participants gain a little, including the country releasing the stocks which nevertheless carries the full cost.

A more general question, however, is why should governments hold stocks rather than the private sector? The more governments hold stocks the less incentive for the private sector to hold them – since the incentive is to benefit from high prices when they occur but which may not occur if governments release stocks. The logic for government involvement is that the private sector has been providing insufficient stock levels under competitive pressure. Yet, if it does not pay the private sector to hold adequate stocks, however, it will pay even less when the government also holds stocks.

As an alternative means of hedging against the risk of price variations, futures markets have been used by Mexico and the State of Texas. Although these were exporters, there is no reason why smaller countries at least could not use such mechanisms for offsetting import price variability (Daniel, 2001).

Diversification of suppliers

Greater energy security has been sought through increased diversification of oil and gas supply sources – and locking in control of external oil and gas sources. The major Asian oil importers have been active in attempting to diversify their import sources of oil and gas and to tie down the security of supply from their suppliers with investments linked to supply options within their investment contracts. For example, China has invested in production and exploration in the Middle East, the Americas, Africa, South-East Asia and, closer to home, in Russia and Central Asia. Like Japan and South Korea, China is proposing major oil and gas pipeline investments in Russia. Japan has supported investments in overseas energy projects in areas close to the region, such as Indonesia, Australia and Russia, as well as further afield in the Middle East. South Korea has similarly pursued supply diversification and supply security efforts internationally, with exploration and production interests in 13 countries.

Supply diversification has benefits in both short- and long-term situations in spreading risk – but it also has costs. Buying into existing operations can lead to offering unduly high prices unless information is accurate – and some analysts have suggested that for some of China's purchases the 'scramble' effect may have led to unduly high prices being offered.[15] Dependence on market supplies will still remain the major component of imports. Even with its major diversification efforts, China's 'equity oil' is unlikely to account for more than a relatively small percentage (perhaps 15 per cent) of current oil import requirements. Moreover, 'equity oil' does not protect against sea-lane interdiction or price hikes.

US concerns about China's quest to lock up energy supplies, rather than participate in energy markets are probably exaggerated (Hadley, 2006). Oil from tied links will reduce China's purchases from elsewhere in the oil market – with an overall neutral effect on world oil prices.[16]

Diversification of fuels

Global oil reserves tend to be relatively concentrated and to be located geographically in high-risk areas. Diversification of fuels, can contribute to energy security through switching from oil (and coal) to gas, and to increased use of nuclear power and renewable energy sources. The shift to alternative fuels has been increasingly to natural gas and exemplifies an environmental improvement in terms of greenhouse gas and other emissions, especially with respect to coal. However, fuel switching may not always be environmentally benign, as with processing of tar sands or liquefaction of coal. Similarly, China's proposed addition of another 30 nuclear power plants to its current seven will require substantially increased imports of uranium and also raises questions about the long-term disposal of nuclear waste.

Asia's demand for natural gas is increasing more rapidly than in other regions. Japan and South Korea are presently the major importers but China and India will become major importers substituting gas primarily for coal in power generation and in industry. Since these two countries have large domestic coal resources, the switch to gas does not completely enhance their energy security as it involves increased import dependence and potentially greater vulnerability to supply interruptions. Nevertheless, it does diversify import sources, particularly when it reduces the transport distances involved.

Gas supply security is, at present, more limited than for oil because of the extent to which trade is dependent upon fixed pipelines. As noted earlier, however, the role of LNG is growing rapidly, thereby increasing flexibility.

Domestic production

Asian countries with some domestic energy resources are increasing development efforts; for coal, efforts to improve exploitation and distribution efficiency including using technologies such as clean coal and coal liquefaction, have grown. Apart from gas, where gas liquids are becoming more important, these 'alternative' fuel sources do not contribute much to meeting the demand for transport fuels. Despite considerable attention to renewable energies, notably in China, with growth mainly in biomass and wind energy, projection of their potential contribution suggests that it will remain at about 2 per cent of a higher level of global demand by 2030. The contribution of hydroelectricity is unlikely to change significantly.

Information and transparency

Although past energy security efforts commonly involved a greater perceived role for governments to control and regulate the industry, much of the international interest is now directed to making the market work more effectively, i.e. more flexibly, and so contributing to energy security. The removal of market impediments, including what the IEA and the IMF see as a serious problem of inadequate and inaccurate market information, and often arbitrary and unpredictable controls and regulations in oil-producing countries, is therefore important. The lack of timely and accurate oil market data on demands and supply, trade and stocks is believed to have exacerbated the market situation in 2004, increasing the perception of risk, reducing willingness to invest in new capacity and increasing price volatility (IMF, 2004: 172). Greater information transparency is seen as a priority for improving efficient operation of the oil market. Various efforts through the International Energy Forum, to which Asian countries and regional organizations such as ASEAN and APEC belong, have improved data availability but it is still deficient particularly on oil demand, supply and stocks data.

Increased investment in the energy industry is a major factor in seeking energy security. Seeking more consistent decision-making by transitional economy governments about investments in the energy industries has been an important target of the Energy Charter Treaty of the EU with industry. International interest has been shown in extending it to the Asian region given the regulatory and related problems that exist in Asia.[17]

Good relations with supplying countries

Governmental efforts to ensure good relationships with supplying countries, such as the dialogues between OPEC countries and the EU and APEC countries are important; arrangements with supplying countries to improve the investment climate would include cross investments with importing countries investing in production in the exporting country, where this is permitted, and exporters investing in refinery capacity in importing countries (for example, Saudi investment in Chinese oil refineries and Chinese investment in Saudi exploration and development). Such dialogues could help facilitate further capital flows and more predictable controls and regulations. They could also contribute to another mechanism to reduce longer term price volatility – the maintenance of surge or spare capacity by oil and gas producers. This would involve added costs for producers – but energy producers also benefit from price stability.

Increasing international co-operation

Measures that enhance energy supply security are commonly of a global public good nature. That is, countries other than those taking the measures may gain as well. Other measures may be more effective when done collectively with other countries. For these reasons, it will often pay for countries to act together for various energy security purposes – joint stock management, swap arrangements, data provision, technology development in conservation, energies and fuel switching, and in pipeline and electrical transmission systems.

Reference has already been made to multilateral co-operation among oil-importing economies to reduce oil consumption, especially through policy intervention promoting fuel efficiency and addressing the national and global external diseconomies of oil consumption, notably global warming. It was noted that such international co-operation could be justified by reference to 'global public good' arguments, as well as being a counter to oil producers' market power. Constructive co-operation is apparent in a Japan–China agreement that encompasses China's imports of conservation techniques and Japan's imports of China's coal.[18]

Regional countries, in the APEC context, have been active in seeking to pursue energy security measures through the APEC Energy Security Initiative. Under this initiative they are considering a variety of measures, both short- and long-term, under five headings: energy supply disruptions; energy investment; energy efficiency in use; diversification; and technology innovation.

Conclusion

We noted earlier that the current high prices can be attributed to a surging demand, a long period of under-investment in energy supply, processing and distribution as well as some speculative elements. In due course, the depletion of the fossil fuel base will constrain production but that seems to be some time off. Although the global and regional economies have coped well so far with oil price increases, the consequences of sustained higher oil prices are likely to become serious for the global and regional economies.

Compared with Europe and North America, Asian countries are less able to cope with such high prices since oil is becoming a larger component of economic activity and of their import bills and the economic policy structure is usually less developed. Issues of energy security are particularly sensitive in Asia given their rapid prospective economic growth and consequent future energy needs and their less well-developed links with the energy supply industry than in the West.

Developments in energy markets in the future face critical uncertainties, notably about the scope for further discovery of conventional energy sources; the levels of investment in all stages of the industry on which future output depends, and the technical capacity of existing and new wells to expand production at rates consistent with demand growth expectations. For the Asian states, various policies can be introduced or expanded to improve energy security in the face of an uncertain future for global energy supplies and prices. Of these, the least cost measures include conservation and increasing energy efficiencies, which also have environmental benefits. The various energy security aspects need to be brought together at national levels to ensure coherence in national energy policies in Asian economies. Regional as well as international energy co-operation would also help in developing more effective policies to reduce energy insecurity in Asia.

Notes

1 IEA (2004b: 70); references to IEA, US Energy Information Administration (EIA) and IMF projections are to their 'reference' scenarios unless otherwise stated.

2 IEA (2004b: 54); both the IMF and the IEA studies referred to the effects on global GDP of sustained oil price increases. At the high oil price of $50/barrel, at which the 10 per cent and $5/barrel shocks are equal, modelled impacts on GDP with respect to the IMF and IEA studies over the 2-year period were to reduce GDP by 0.6 per cent and 0.8 per cent, respectively. If we consider the effects of a 10 per cent shock at lower prices in both models, predicted effects on GDP in the IMF case become significantly lower than in the IEA model.

3 IMF (2005b: 157); in September 2005, the IMF reduced its global economic growth forecast from 4.4 per cent to 4.3 per cent due to higher energy prices. *Bloomberg*, 26 September 2005.

4 The ADB's membership, to which its data relate, is wider than some other regional definitions, consisting of 63 countries including many Pacific Island states and several of the Central Asian states. Of the latter, Kazakhstan with respect to oil and Turkmenistan with respect to gas, are expected to become increasingly important oil and gas exporters, especially to the Asia–Pacific.

5 IEA reference scenario. For the EIA, the OPEC share varies in the range of 40 Mb/d to 73 Mb/d by 2025 corresponding to oil prices of US$48/barrel and US$21/barrel (EIA, 2005c).

6 In IEA models, prices are not forecasts but judgements about the prices needed to balance future supply and demand.

7 The objectives of some neoconservative analysts were said to be to 'break OPEC', or more specifically, to break its supposed 'market power' by introducing a hypothetically US-aligned Iraq as a major oil supplier achieved by privatising its national oil company and opening entry for international oil companies. For US policy, there was a more serious problem with this policy, apparently since abandoned as impractical. This was that a 'low oil price scenario', however achieved, is incompatible with the United States' preferred approach to energy security, that of 'supply diversity' and maximum encouragement to both IOCs and non-OPEC production, both of which require high oil prices. See Cohen and O'Driscoll (2002); Palast (2005: 74–76).

8 'Ali al-Naimi calls for building more refineries in consumer countries', *Alexander's Gas and Oil Connections*, 10(3), 6 July 2005.

9 The price spread between light and heavy oils – the latter making up a significant proportion of OPEC's marginal production – was around US$8–9 in September, reflecting the particular refinery bottleneck (IMF, 2005b: 58).

10 Reuters, 'Goldman Sachs: Oil Could Spike to $105'. *Energy Bulletin*, 31 March 2005; *Petroleum Economist*, May 2005, (lead article).

11 Private companies will also seek to reduce risks in various ways: stockholding; use of long-term price contracts; futures markets; joint ventures; market diversification etc. We concentrate here, however, on governmental activity. Nevertheless, how far the private sector employs mechanisms to reduce risks reduces the need for governmental action. In reverse, governmental action e.g. on stock holding may reduce the incentives for private sector activity.

12 Larsson (2006: 262–266); Larsson notes that Russia's use of such leverage is quite limited with respect to other countries. Nevertheless, its dispute with Ukraine over gas prices may have created an atmosphere of uncertainty.

13 Proven reserves have been defined as those with a probability of at least 90 per cent that the estimated volumes can be produced profitably. Differences in application of criteria even for proven reserves are illustrated in (i) Kazakhstan, where expert sources differ widely (BP–Amoco: 40 billion barrels) and (*Oil and Gas Journal*: 9 billion barrels); (ii) Canadian tar sands (174 billion barrels) – included by *The Oil and Gas Journal* but excluded by BP–Amoco; and (iii) in challenges to Saudi Arabian proven reserves as underestimates by some analysts and overestimates, most recently by Matthew Simmons (2005).

14 'Follow the Chopsticks'. *New York Times* (editorial), 25 March 2006.

15 Unless (despite the earlier comment) we accept that a Chinese perception of much higher future oil prices is accurate.
16 A point now acknowledged by the US Department of Energy (2006).
17 Originally designed to provide a kind of WTO framework for energy relations between Western Europe and emerging Eastern European states, particular interest has been shown in North-East Asia in a similar regional treaty.
18 *Xinhua News Agency*, Beijing, 5 December 2005.

11 The environmental effects of energy competition in the Asia–Pacific

Aynsley Kellow

The recent announcement of the formation of an Asia–Pacific Partnership on Clean Development and Climate (APPCDC) was of momentous importance for the future of the region – an importance which would be too easy to underestimate. It has great potential to grow to incorporate further parties, and sets the scene for the future relationship between energy and environment in the region.

If the Kyoto Protocol to the Framework Convention on Climate Change (FCCC) cannot yet be pronounced dead, it is at the very least not too healthy. It could, perhaps, be described as 'undead', rather than demonstrating full vital signs. Nobody appears quite prepared to drive a stake through its heart. Or perhaps, in the terminology of a famous Monty Python sketch, it is a dead parrot, nailed to its perch by those trying to sell it to a reluctant public.

Kyoto's death knell had been sounded in the weeks before the announcement of APPCDC (July 2005) in the communiqué on climate change issued by the G-8 leaders at Gleneagles the previous May. Significantly, while the communiqué sought to save face for those leaders committed to Kyoto, it stated that the FCCC (not Kyoto) provided the appropriate framework for the future for addressing the problem of climate change. The statement merely noted that those who had made commitments under Kyoto would honour them, but the three problems with this were, first, that it was apparent to most that it was likely many of these countries will fail to meet their targets; second, even if they did, there would be negligible benefits for the considerable costs involved; and finally, it was clear that Kyoto did not provide a productive basis for action beyond 2012.

To take these in turn, even with trading under the 'European Bubble' (or Burden Sharing Agreement (BSA)) the windfall reductions of the United Kingdom (resulting from the 'dash to gas') and Germany (resulting from the economic collapse of the former East German economy), Europe looked as if it would need to purchase emission entitlements from Russia to meet its collective commitments. Japan and Canada were also experiencing difficulties, and New Zealand had found that, rather than an economic bonanza, Kyoto was likely to cost it money in purchased entitlements.

Second, reliance on trading, both under the European BSA and more widely, meant that there would be an even more modest reduction in greenhouse gas (GHG) emissions than had been anticipated as a result of Kyoto. Kyoto would make an almost unnoticeable effect on GHG levels at 2100, and it was widely justified on the basis that it was merely a first step. It is now also widely regarded as having been a first step in the wrong direction. It imposed a 'cap-and-trade' system which would do little to bring about any actual reductions in GHG emissions, since those who had not already made largely involuntary progress post-1990 could meet their targets by buying what many had long regarded as 'hot air' from the Russians – reductions attributable to their post-communist economic collapse. Russia was largely happy to sell in the short-term, but saw itself as needing entitlements in the longer term as its economy developed. Meantime, it was happy to trade hot air for EU support for WTO membership and investment in the development of its gas sector.

Despite substantial European investments in technology like wind energy, the substance was that Kyoto let Europe off lightly, and would produce virtually unnoticeable effects in GHG mitigation. It was most advantageous to the two nations – Germany and the United Kingdom – which bore the greatest responsibility for the historical accumulation of GHGs in the atmosphere. Since the residence time of GHGs like CO_2 is of the order of 100 years, Kyoto was wrong to treat the problem as if it were one of flows, rather than stocks, and the high per capita present day emitters were expected by Kyoto to do more to address the problem than those who had disproportionately contributed to its cause. This unfairness of Kyoto was exacerbated by the selection of 1990 as a base year against which emission reduction targets would be specified, because it so advantaged those (the UK, Germany, Russia) which had fortuitously seen GHG reductions occur after that year, especially in relation to those (such as Japan) which had undertaken significant measures to improve energy efficiency prior to 1990.

As much as any other reason, this meant Kyoto was a poor model for future progress in addressing climate change. While it imposed no immediate costs on developing countries, it would be ineffective unless it evolved so as to bring about at least a slowing in future GHG emissions as such countries industrialised. Kyoto did promise, through mechanisms such as trading, activities implemented jointly and the clean development mechanism, incentives for participation, but these would be more than offset by any future imposition of constraints on their economies. If Kyoto required little of its strongest supporters, why should developing countries take on commitments which might limit their development?

Beyond Kyoto: Europe versus Asia–Pacific

It was therefore apparent at the Conference of the Parties to FCCC meeting at Buenos Aires in December 2004 that Kyoto had effectively been abandoned by both the US and allies and the key developing countries, China and India. If there was to be progress in addressing climate change, it was clear from then on that it would not be with some kind of 'Son of Kyoto'. Russia can therefore be seen to have conducted a masterful negotiation to extract maximum benefit from the opportunity presented by Kyoto, and to ratify just before Kyoto was effectively put on life support – effectively dead before it even entered into force in February 2005.

It has to be said that this result represents a substantial achievement for the Australian government. While various accounts of the process by which APPCDC was developed have surfaced, and the prominent role of Australia has been acknowledged, it has generally been suggested that this role was essentially as some kind of deputy to the Bush administration in the USA. This interpretation might be correct as far as it goes, but it does not give due recognition to the almost lone role played by Australia in the 3 years after the negotiation of Kyoto and before the election of George W. Bush, when Australia was largely isolated on climate change, with the government responding to vigorous lobbying by industry which was keen to point out the hazards of a policy instrument which would facilitate substantial 'carbon leakage' to other economies in the region, potentially limiting Australia's ability to make the most of the current substantial resources boom led by the rapid industrialisation of China.

It is important to recall that not only was Bill Clinton in the White House, but Vice-President Al Gore was playing a leading role in climate change negotiations, and the Clinton White House was among those making strong and vigorous representations to the effect that Australia should ratify Kyoto. True, the Byrd–Hagel Resolution in 1997 had made it quite clear that ratification of Kyoto by US Senate was highly improbable, but the Clinton Administration was clearly playing 'two-level games', harvesting the green vote by championing strongly an agreement it would not have to live with. Word at the time was that a Gore White House would not have introduced Kyoto into the Senate for ratification until a second term had been secured – a variation on the Not In My (First) Term of Office (NIMTO) strategy.

Australia, the US and other members of the 'Umbrella Group' had all been more reluctant than the EU to ratify Kyoto, essentially because the costs and benefits involved were so markedly different. Regardless of whether the economic modelling on which they were based was accurate and reliable, the two groups stated publicly figures for the economic impact upon each that were two orders of magnitude apart. The Australian recalcitrance on this issue was supported by powerful interests in both business and labour. While the ALP supported ratification, it is important to note that opposition leader Kim Beasley was warned against ratification in 2001 by the CFMEU. Whether a Beasley government *would* have ratified in the face of these interests will never be known, but the Howard government toughed it out, despite finding itself substantially isolated for 3 years.

Kyoto was in so many ways a Eurocentric agreement. Not only did it not suit the interests of resource-extractive economies like that of Australia, it worked to the relative advantage of Europe. It was not surprising that a country whose largest export was coal, and whose economic output was comprised so much of mining and metals should resist an agreement which suited so well those who are among the final consumers of such products, but the important lesson of Kyoto is that it followed the wrong negotiating process. The point is not that either Europe or Australia is to be criticised for following its interests, but that that they should do so was inevitable.

The negotiating process on climate change was modelled quite explicitly (by UNEP's Mustafa Tolba) on what was understood to have been a successful process in responding to the problem of ozone depletion. That process was much studied, and has widely been described as involving the emergence of an epistemic consensus, a confluence of science and normative beliefs which proved robust to competing or alternative views. But many analyses of the process by which the Montreal Protocol was negotiated ignored the central role of interests in that process, especially the powerful interests of US chemical companies which owned intellectual property and saw advantage in a CFC phase-out, and who therefore supported a positive US role in the negotiations.

Climate change was always going to be different. The science is inevitably less determinate, and the interests more substantial and vexsome. Some key players saw advantage in the issue. Then British Prime Minister Margaret Thatcher found in it a reason to close uncompetitive coal mines, and there is no clearer indication of the divide between the EU and the Umbrella Group than in the fact that the cost of steaming coal in Europe (the worst GHG-emitting fossil fuel) was four times the price in the USA, Canada and Australia. German Chancellor Helmut Kohl found in it a means of supporting 'Green Keynesianism' and the German nuclear program after 1986 brought both a collapse in the oil price which underwrote the former and the Chernobyl accident which jeopardised the latter. The Social Democrats and Greens at the 1987 election were proposing nuclear phase-out and expansion of coal, so climate change as an issue compromised the policies of Kohl's opponents. More generally, the issue has also bolstered European energy security policies, which have included such measures as highly unpopular fuel taxes, with real petrol prices at the pump typically double those in Australia, despite a policy of world parity pricing in the latter.

Such concerns with energy security also underpin the continued payment of substantial coal subsidies in the EU. Australian coal producers a decade ago harboured hopes of improved market access to Europe as an outcome of the Uruguay Round of trade negotiations, and it has often been pointed out that coal trade reform would produce both environmental and economic benefits for all, as the coal exported by producers such as Australia is both cleaner and cheaper than that being subsidised in Europe.

A new climate partnership

This leads on the question of what the effective demise of Kyoto and emergence of APPCDC means, especially for environmental aspects of energy security in the Asia–Pacific region. First I shall deal with the implications of APPCDC for climate policy, because this will inevitably frame more general matters.

The first point to note is that APPCDC has the potential to be a marked improvement over Kyoto in addressing climate change. Kyoto was too much a case of 'garbage can' decision-making – of preferred solutions finding problems which justified them. Those most enthusiastic about Kyoto have been those harmed least, or those who saw potential advantage in it. A good example of the latter is Canada. If you are in a free-trade agreement (with interconnected electricity grids) with the largest GHG emitter and have substantial hydro-electric resources and both low-cost uranium and CANDU reactor technology, instruments requiring decarbonisation are an opportunity, not a threat. Any regulation creates both advantages and disadvantages.

And while Kyoto was superficially attractive, it was flawed. By this statement, I mean that 'cap-and-trade' approaches are by far the most desirable instrument for regulating environmental pollution. They allow abatement to occur at lowest marginal cost and minimise distortions. But the Kyoto cap was neither 'one size fits all' nor required to be worn by all. The true test of international climate policy was always going to be what kind of means would be adopted by those whose future emissions (as they industrialised) had the potential to render meaningless the efforts of the FCCC Annexe I Parties to mitigate their emissions.

There is an attraction in regulation – in sticks, rather than carrots – but it should not be supposed that a carrot-based approach such as APPCDC appears to be is without merit. Regulation is sometimes redundant, and has been credited in the past with successful outcomes for which the relationship between cause and effect is illusory (despite becoming firmly entrenched in the collective consciousness). One example of this which is germane to APPCDC is the effect of the Clean Air Act in the UK. Many attribute the end of urban smogs in Britain to the effectiveness of the Clean Air Act; even quite recently, Michael Grubb cited the Act as an example of successful regulation in criticising Bjorn Lomborg's *The Sceptical Environmentalist*. But the impact of the legislation was negligible, with the same effects attributed to the Clean Air Zones being declared under the Act manifesting themselves in areas where no declaration was made. The improvement in urban air quality resulted overwhelmingly from socio-economic factors and technological improvement: consumer switching away from coal fires to forms of heating which were more convenient and increasingly affordable (Scarrow, 1972: 261–282; Auliciems & Burton, 1973: 1063–1070). Urban air quality in Britain therefore provides a positive portent for the possible value of APPCDC: improvements *can* result from rising affluence and technological improvements.

On the other hand, regulation of the wrong kind can actually exacerbate problems. A stark example of this was the 1990 Clean Air Amendments in the USA. These required specified percentage removal of sulphur oxides from flue gases, which could be more readily achieved if the plant was burning high-sulphur coal. The legislation also encouraged owners to keep older plants in service, thus slowing investment in newer, cleaner plants, so that air pollution was worse than if there had been no policy in place. Like Kyoto, the policy was the result of a coalition between environmentalists and those with less noble motives: Appalachian coal producers and their unionised workforce, whose high-sulphur coal was advantaged vis-à-vis low-sulphur coal from non-unionised western coal producers.

These examples show that the demise of Kyoto and the emergence of APPCDC need not give rise to concern about whether there will be an effective international response to climate change. This is especially so if we accept that climate change presents a classic case of decision-making under uncertainty, where the dominant problem construction thus far suits numerous interests, but does not necessarily favour the best policy responses. Climate change science is riddled with uncertainty, and most of what we 'know' about the future is virtual in nature. Economic scenarios drive emissions scenarios which in turn drive climate models, which are inevitably judged by the extent to which they agree with other models and the (shorter than desirable) series of data used to tune them. When dealing with coupled, non-linear systems, we are inevitably confronted with a plethora of 'coulds' and 'mights': how one responds to such science depends upon how various problem sets and their solutions intersect with one's interests. Until models can be tested against a sufficiently long series of observational data, many of these uncertainties will remain and interpretations of the problem and solutions to it will be deeply affected by interests and 'garbage can' responses.

The uncertainty inherent in the science will continue to limit the possibility of an epistemic consensus emerging to overcome divergent interests. Significantly, Peter Haas predicted as much while the climate change negotiations were in their infancy. This makes an approach focused on adaptation, development assistance, increasing affluence and investment in research and development much more promising if developing countries are to be brought on board, and their future emissions are to be constrained without threatening their economic development. Partnerships such as APPCDC are undoubtedly less appealing to those focused more on the appearance of action, rather than the possibility of voluntary commitments, but (as the policy instruments given as examples above indicate) there is no necessary reason why they should be less effective, and might even be more so. After all, as US Deputy Secretary of State Bob Zoellick was quick to point out at the press conference in Vientiane at which APPCDC was announced, in the first 3 years of the Bush Administration US GHG emissions reduced marginally, while those of the EU 15 rose by 3.6 per cent and those of the EU 25 by 3.4 per cent.

One fundamental problem with Kyoto has been a focus on decarbonisation first and foremost. It applies to a basket of GHGs other than CO_2, and all of these are much more powerful in their effects than CO_2. It is the sheer scale of CO_2 emissions which has resulted in much of the discourse simply ignoring these other gases, as well as a whole raft of contributing factors – or, perhaps, more accurately, that a focus on CO_2 has also suited some interests. There is an appealing simplicity in targeting the largest single agent of greenhouse forcing, but it has not necessarily been the wisest policy approach to do so, and the APPCDC reflects a recognition that other components of the problem could be mitigated either more readily (in a technical sense) or more economically.

The APPCDC reflects the science behind what is known as the Hansen Alternative Scenario (Hansen et al., 2000: 9875–9880), which acknowledges that while CO_2 is the largest forcing, it does not dwarf the others. Indeed, Hansen et *al.* (2000: 9876) contend that because fossil fuel consumption has been accompanied by both CO_2 emissions and cooling aerosols, the non-CO_2 GHGs 'have been the primary drive for climate change in the past century'. Hansen *et al.* argued for a focus on drivers of climate change (such as black soot and tropospheric ozone) which could be mitigated more readily and more cheaply, while noting that 'Investments in technology to improve energy efficiency and develop nonfossil energy sources are also needed to slow the growth of CO_2 emissions and expand future policy options' (Hansen *et al.*, 2000: 9879). The attraction of focusing on forcings such as black soot is that it also provides significant incidental benefits such as improvements in mortality from emissions from diesel engines and coal-fired power plants, and from combustion of biofuels (significant in India and other parts of Asia, where indoor air pollution is perhaps the most serious environmental problem – especially for women and children).

This is the scientific basis of APPCDC, which builds upon the earlier Australian/US Climate Action Partnership concluded in February 2002. The six founding members of APPCDC account for 49 per cent of global GDP, 48 per cent of global energy consumption, 48 per cent of global GHG emissions and 45 per cent of population. Perhaps more importantly, by including China and India, it addresses the most significant sources of *future* GHG emissions. It will seek to foster research and investment aimed at cleaner, more efficient energy utilisation in the Asia–Pacific region, including renewable energy and energy efficiency, remote area power supplies, liquefied natural gases (LNGs), methane capture and use, clean coal, nuclear power, advanced transportation, agriculture and forestry. Success is not guaranteed, but others have already expressed an interest in joining a framework which surpasses Kyoto in its coverage.

Clearly, APPCDC sets the environmental framework for future energy development in the region. Together with economics and energy security considerations, it sets the scene for future energy policy in the region. Not only will APPCDC affect the energy policy orientations of governments in the region, but it will impact on both aid and investment polices of its members and ultimately, one would expect, on the investment policies of the World Bank. The World Bank has been lobbied heavily by NGOs opposed to hydro-electric development, and its Materials Review came close to resulting in restrictions on investments in coal energy. APPCDC is already too large to be ignored, and it is likely that more parties will join. It is hard to see that it would not impact upon investment decisions at both the national and international levels.

Energy resources in environmental context

We can now move to consider how energy competition in the Asia–Pacific is likely to develop in the context of APPCDC, which will accentuate energy sources which make sense to the region not just in terms of environmental issues like climate change, but also in the context or the development trajectories of states, and considerations such as energy security and potential conflicts.[1] We look first at conventional sources and then at alternative energy.

Conventional sources

Part of the future emphasis will be on investment in 'clean coal' and other technology, especially technology to remove CO_2 from fossil fuel combustion plants. This is especially important given the resource endowment of the region, with the bulk of the world's proven high-rank coal reserves lying in North America, Asia, and Oceania. Both North America and Asia have over 25 per cent each of total reserves. While Europe still has substantial reserves, these are mostly of lower rank coals. Reserve estimates tell only part of the story, the other being price, and the cost disparity of steaming coal in Europe and producers such as the US, Australia, South Africa and Indonesia reached a factor of four by the time Kyoto was being negotiated. Given the fact that coal generates the most CO_2 per joule of energy, there is little surprise in the reluctance of the US and Australia to ratify, and the attraction of APPCDC to them and other states in the Asia–Pacific.

Europe and the Asia–Pacific have widely divergent views of coal, and there is little coal industry left in countries like the United Kingdom and France. Germany's output of both hard coal and lignite fell by half over the 1990s. Its hard coal lies in deep seams (over 900 m deep) and it is thus expensive to mine. It has reserves totalling over 300 years' production, but it is too expensive to be internationally competitive and is heavily subsidised. This is also the case with the UK, which has only 40 years' production. But (relatively speaking) there is abundant cheap coal in the Asia–Pacific region: Australia has 270 years' production, the US 250 years, China 111 years, Canada 90 years, India 268, Indonesia 76, Pakistan 686, Russia 629, and Thailand 69 (WEC, 2005). Given this resource distribution, it is clear that coal will lie near the centre of the economic development of the region, in a way that is markedly different from that of Europe. Clearly, clean coal technology is of enormous interest in the region.

The collection of CO_2 from large-scale combustion plants was already happening on a commercial basis in the US when Kyoto was negotiated. CO_2 is a commercial product; there is a commercial well near Millicent in South Australia. CO_2 can be removed from power station emissions by scrubbing with monoethanolamine (MEA), but much of the cost of MEA scrubbing is involved in the flue gas desulphurisation (FGD) which must be undertaken first. The cost penalty is not likely to be so large with low-sulphur Australian coals. The problems lie not so much in collection, but in disposal, once there is an abundance of CO_2. It is already used to inject into oil fields to enhance production, and the use of such oil or gas reservoirs for sequestration is possible. Deep ocean dumping is attractive in capacity terms, but while it is technically feasible in theory, cost is likely to be prohibitive.

Even without removal of CO_2 from flue gases, the most modern technology for coal-fired power generation offers marginal improvements in emissions over traditional technology, and more substantial improvements over most existing plants. Super-Critical Steam technology offers about a 4 per cent reduction in CO_2 emissions over the best Sub-Critical Steam plant (and slightly more over most plants currently in use). Combined Cycle Advanced Frame GT using natural gas offers a slightly more than 50 per cent reduction over Sub-Critical Steam. Using coal in Integrated Gasification Combined Cycle can result in greater efficiencies, but its potential contribution to climate change comes from its potential for CO_2 capture and sequestration.

Using oxygen rather than air in a coal gasifier produces CO_2 in a concentrated gas stream, which can be captured more easily and at lower costs for ultimate disposition in various sequestration approaches. Gasification also promises improvements in other environmental impacts from coal, such as particulates, SOx and NOx. Emissions of CO_2 are higher from coal than from either natural gas or distillate oil-fired combined cycle units, and the competitive price of natural gas makes it attractive in a greenhouse environment as the easiest way of achieving substantial decarbonisation at relatively low cost. As gas and oil are frequently co-products, this means that both will be significant in the energy future of the region. The future is likely to see overlapping phases of development: a 'dash to gas' similar to that which after 1990 made Kyoto so advantageous for Europe, followed by increasing penetration of clean coal technology.

Gas is already becoming an important regional energy source. China's consumption of coal in 1999 decreased, but it increased its natural gas consumption by 10.9 per cent over 1998. In the Asia–Pacific region, consumption of natural gas increased by 6.5 per cent in 1999. It is anticipated that a fairly significant portion of future energy demand in the region will be met by natural gas. The export trade in natural gas (especially as LNG) is important: of the 485 billion cubic metres of gas traded internationally in 1999, about 25 per cent or 124 billion cubic metres was transported in the form of LNG, 75 per cent of which was shipped to the Asia–Pacific.

Unlike coal, of which there are substantial reserves in the region, the uneven distribution of oil and gas reserves could give rise to future concerns over security of supply and tensions over new and potential resources. Concerns over secure access to resources are common to many countries, given the concentration of petrochemical resources in the Middle East, and freedom of navigation is of obvious importance, but the emergence of new technologies allowing deep ocean exploration and extraction during the 1990s will undoubtedly heighten tensions over maritime and other territorial disputes. We have seen some of these tensions over the Timor Gap, but disputes in the South China Sea (over the Spratleys, for example) are potentially more serious.

The place of gas and coal in the region's future electricity supply is strongly linked to its oil future. It may well be that, (as former Saudi oil minister Sheikh Zaki Yamani so colourfully put it) just as the stone age was not ended by a shortage of stones, the age of oil will not be ended by a shortage of oil. But the location of reserves is the basis of concerns over energy security, and the great challenge for countries which lack the enormous reserves to be found in the Middle East is to find energy resources which can replace the role played by liquid fossil fuels in transportation. Gas has some potential, but it seems the days of the internal combustion engine are numbered and hydrogen fuel cells are the way of the future. But the hydrogen economy requires hydrogen, and hydrolysis requires electricity, and this means that other ways of generating electricity will be important in the region (as elsewhere).

Constraints on carbon emissions will also work to advantage development of the region's substantial hydro-electric resources, and it has the largest potential of any region. Hydro development has been opposed by the California-based NGO, the International Rivers Network (IRN), which has enjoyed some success in lobbying the World Bank and other agencies against hydro development. Climate change poses questions for such environmental opposition, and undoubtedly alters the balance of the environmental equation against the IRN.

The hydro-electric resources of China alone are enormous: its gross theoretical potential approaches 6,000 TWh/year, while its economically feasible potential has been assessed as some 290,000 MW (1,260 TWh/year) – the most abundant in any country in the world, let alone the region. Current hydro output exceeds 200 TWh/year, contributing about 17 per cent to the republic's electricity generation. There is potential, therefore, for a six-fold increase in hydro generation, and this will become more attractive in a carbon-constrained world. There is also considerable hydro potential in India, Indonesia, Malaysia, Nepal, Pakistan, Russia and Vietnam.

Hydro-electric development is a potential source of both co-operation and conflict. Several river basins (most notably the Mekong) transect national boundaries, and development could interfere with flow characteristics, and even result in abstraction. Both have consequences for downstream countries, and present potential points of conflict which require management. The potential for co-operation is not quite so obvious, but the financial scale of potential development in countries such as Nepal (an important factor in its development prospects) is likely to make development dependent on power export contracts with another state, such as India.

Nuclear energy raises a whole different set of environmental and security challenges which it is not necessary to revisit here, but the nuclear industry has had a renaissance thanks to climate change, with even environmentalists calling for a re-opening of the debate. The nuclear path is already being followed in the region, and this is likely to continue.

China is a relative newcomer to nuclear energy, and its nuclear weapons program preceded its energy program, not the other way around. Its first nuclear power plant, a 279 MWe PWR at Qinshan, near Shanghai, was commissioned in December 1991. There followed two larger PWRs (each 944 MWe) at Daya Bay (Guangdong province) in 1993–1994. By the end of 1999, China's nuclear generating capacity was 2,167 MWe, with output from these three units providing only 1.2 per cent of electricity generation. However, seven more nuclear units were under construction at the end of 1999, with a capacity of about 5.4 GWe.

At that time, India had 11 reactor units in operation, with an aggregate net generating capacity of 1,897 MWe. Nuclear sources accounted for 2.7 per cent of total electricity generation in 1999. Another three plants came into operation during 2000, and construction of a further two were under construction at Tarapur with construction beginning on the first of two Russian-designed reactors at

Kudankulam in Tamil Nadu in 2001. A further 14 were planned. India's long-term objective for nuclear capacity is 20,000 MWe (gross) by 2020: in order to achieve this goal, it plans to develop fast breeder reactors and to use its substantial indigenous reserves of thorium.

For the two largest countries in the region, therefore, the nuclear option is near the forefront of its energy planning, and the ability to source energy resources from domestic sources is clearly an important factor in their decision-making. The Republic of Korea has a sizable commitment to nuclear energy, with a target of 33 per cent of total generating capacity, by 2015. This proportion is already exceeded in Japan, and by 2010 there are expected to be 64 nuclear reactors in operation, with a total gross capacity of 60,316 MWe. Having few indigenous energy resources, nuclear energy is important for Japan as a stable supply of electricity, and while it is politically contentious, it is important to remember that Japan is a party to Kyoto, and is likely to pursue both emissions trading and the nuclear option, as well as pursuing technologically based options under APPCDC. There are also more modest nuclear plans in Bangladesh, Indonesia, Pakistan, the Philippines, and Vietnam.

The transfer of nuclear technology to Pakistan by China and to India by the former Soviet Union, and the subsequent development of nuclear weapons by both underscores the strategic importance of this technology, which has been made more attractive by the climate change issue. The recent tensions over the nuclear program of the People's Democratic Republic of Korea, and the substantial existing and planned nuclear program of Taiwan underscores the inter-relationships between energy security and competitiveness and strategic considerations in the region. But the nuclear weapons genie is well and truly out of the bottle in the region, and the challenge is to manage the problem; it is already too late to prevent proliferation.

Alternative energy sources

Thus far, we have overlooked the various renewable and alternative energy resources and what place they might have in the region. They will undoubtedly have a significant place in meeting the energy needs of the Asia–Pacific, but it will be a limited role and one which is subject to various constraints.

The climate change issue certainly enhances the attractiveness of alternative energy sources such as wind, solar and biomass. Globally, both wind and solar have been expanding rapidly (albeit off very low bases), but they suffer from both economic and environmental constraints, and it is unlikely that they can play a role in the region which is anything other than minor.

There are two major environmental concerns with wind energy: the land lost to the wind farm, and threat to local bird populations from the turbines. The land loss is required for the wind farm, and the need for access roads capable of taking cranes necessary for servicing, but there is restricted opportunity for other uses around the wind farm due to noise and safety. Access for maintenance is required, including access for heavy mobile cranes. While efficiency gains have improved their economics, the intermittent nature of the energy source makes costly back-up plants necessary. Power is typically available from 25 to 75 per cent of the time depending on wind conditions; they produce little under calm conditions and must be feathered in strong winds. Even if we ignore this limitation, the scale of energy demands of rapidly developing economies in the region means that wind will play only a limited role. As one analyst recently pointed out, to replace one nuclear station in France, it would have to place a wind generator every 100 m along its entire coastline. As wind farms become more common on a larger scale, opposition to them is growing, and it is unlikely that they can be constructed in sufficient numbers to play anything other than a minor role, even leaving aside marginal economics (and they are the closest of all the alternative generation sources to being competitive).

There are, nevertheless, wind energy programs in the region. The Chinese government has a stated goal of 3,000 MW of installed wind power by 2010, but only 25 MW capacity was added during 1999, bringing the end-year figure to 253 MW (against a 2000 target of 1,000 MW). It has been stated that this slow roll-out has been due in part to government insistence on bilateral donor support for projects, which has kept projects small. This indicates there should be opportunity under APPCDC for investment in wind capacity in China, but it also indicates a scepticism on the part of the Chinese government about the value of wind energy unless supported by donor countries.

There is a more positive attitude to wind energy in India, where a wind energy programme was initiated in 1983–1984. In terms of currently installed wind turbine capacity, India now ranks fifth in the world behind Germany, USA, Denmark and Spain, with 1,081 MW installed at the end of 1999.

Solar energy is technologically appealing, since photovoltaic panels have no moving parts. While costs have come down, however, it is a technology which is competitive only in remote locations in competition with expensive alternatives (such as diesel) or where grid connection is expensive compared to demand (motorway telephones or marine buoys) or impossible (communications satellites). Japan has the most photovoltaic installed capacity in the region with a little over 200 MW, followed by India with 44 MW and China with 9 MW.

The limiting factor for photovoltaic *or* solar thermal is not economic, but physical – in the form of the solar constant. For the Brisbane, Australia, area, the average total horizontal surface solar radiation is approximately 3.5 kilowatt-hours/square meter/day (kWh/m^2-day) in the winter and 7.0 kWh/m^2-day in the summer. This means that enormous land areas are required for megawatt scale electricity generation, which means such generation must be located where land is cheap, such as deserts or building rooftops and the like. With a typical conversion rate of 10 per cent, a surface area of about 2,875,000 m^2 would be required in the winter and about 1,375,000 m^2 in the summer for a nominal generation of about 1,000 MW. Concentrating or tracking mechanisms could potentially reduce this area. However, electric power can only be generated during daylight hours; storage mechanisms such as pumped hydro or back-up plants which are under-utilised are necessary to maintain generation over daily periods.

It was estimated in the mid-1990s that the amount of crop residues amounted to about 3.5–4 billion tonnes annually, with an energy content representing 1.5 billion tonnes oil equivalent (toe). Global energy consumption is about 9 billion toe. Another study estimated that when using only major crops (wheat, rice, maize, barley, and sugar cane), a 25 per cent residue recovery rate could generate about 0.9 billion toe per year. Most of this potential is currently untapped. World generation capacity from agricultural residues (straw, animal slurries, green agricultural wastes) is estimated to be about 4,500 MWt. in 1997. Recoverable residues from forests have been estimated to have an energy potential of about another 0.8 toe per annum, and the potential for energy from dung has been estimated at about 0.5 billion toe.

There is considerable use of fuel sources such as dung in the region, but the environmental consequences of this energy utilisation are serious, with indoor air pollution a serious health hazard for women and children especially. But, as the Hansen Alternative Scenario shows, such energy use (while renewable) contributes substantially to climate forcing through production of carbon soot. The APPCDC has great potential to improve environmental health *and* address climate change by improving utilisation of this energy while promoting development of clean energy sources.

There have been considerable advances in bio-energy utilisation, including: improved integrated biomass gasifier/gas turbine (IBGT) systems for power generation and gas turbine/steam turbine combined cycle (GTCC); circulating fluidised bed (CFB) and integrated gasification combined cycles (IGCC); cogeneration, co-firing; bio-ethanol production; improved techniques for biomass harvesting, transportation and storage; bio-diesel technology; continuous fermentation; improved processes for obtaining ethanol from cellulosic material; better use of by-products; production of methanol and hydrogen from biomass; and fuel cell vehicle technology.

Biomass therefore offers considerable potential for investment in the development of renewable energy, moving it away from the small, often domestic, scale to a scale where greater efficiency is possible and improvements in air quality can be achieved simultaneously with addressing climate change through reducing carbon soot. Understandably, the APPCDC will target such potential.

To wind, solar and biomass we can add numerous other energy technologies: geothermal (both high temperature steam and heat pumps); ocean energy; tide and ocean currents. All will have a role, and the potential for some is considerable, but their immediate prospects are more limited than those surveyed above.

Conclusion

This analysis shows why the APPCDC makes so much sense in the region. It will shape the energy future of the region, and it is highly likely that more and more parties will come on board. There is a neat fit between the energy future of the region and this new approach to climate change. As the Hansen Alternative Scenario shows, it is a more productive approach to focus on least-cost mitigation, measures with co-benefits, and future investment which will direct future development of energy utilisation technologies; these will not only limit future emissions growth, but eventually see the replacement of existing plants with new technologies as old capital plants are retired. APPCDC is a good fit for the Asia–Pacific, just as Kyoto was a good fit for Europe.

It is unlikely, in the process, that Kyoto will be either rejuvenated nor have a stake driven through its heart, but it is likely that it will fade into the past after 2012 and APPCDC will become more prominent. If that is so, we can point perhaps, not just to a new approach to climate policy, but a new model for the development of Multilateral Environmental Agreements (MEAs). Past MEAs have generally disappointed, and I have often argued that the features of their development (conference diplomacy, iterative functionalism, creative ambiguity, double standards provisions, blame and shame tactics by NGOs, and so on) contain the seeds of their failure. In international environmental policy there is much oratory and many good intentions but an insufficiency of good outcomes. It is much as Voltaire said of the French revolutionaries, 'The dilemma for successive generations of those politicians who graduated from oratory to administration was that they owed their own power to precisely the kind of rhetoric that made their subsequent governance impossible'.

APPCDC is an agreement with six parties. It builds on a bilateral agreement between the US and Australia, and it promises to grow to include more parties. The modest scope of the agreement and the limited number of parties have facilitated the development of an approach which, by bringing in China and India (and potentially others) has the potential to be a much more effective policy instrument than Kyoto. Potentiality is not actuality, as Aristotle once observed, but as first steps go, this is one in time with significant 'marchers' and in a promising direction.

Notes

1 Information for this section has been drawn primarily from the World Energy Council (www.worldenergy.org).

12 Energy and human security in the Asia–Pacific: exploring the human security/state security interface

Michael Heazle

As the Asia–Pacific region continues to undergo major economic and political transformation, the challenges and opportunities it now presents are perhaps more influential than ever in shaping the policies of major international actors like the US, the European Union, Russia, Japan, and, of course, China. The geopolitical significance of the region and the import of competition between major state actors (China, Japan, Russia, and the US) are frequently discussed, and in more recent years energy competition and supply has become a hot topic. Such discussions, however, often focus only on how the Asia–Pacific's fast evolving, and increasingly interdependent, political and economic landscape affects inter-state relations in what is essentially a great game of strategy played according to the rules of national interest, military power, and the anarchical system. Comparatively little attention is paid to how transnational threats, like energy shortages, can internally impact on states in ways that the 'great game' perspective has difficulty recognising or understanding.

Indeed, the spiralling costs of oil and the vital energy supplies it is used to produce and distribute constitute a significant threat to both the stability of states and the security and welfare of their respective societies. In all but the most underdeveloped societies, reliable access to affordable, refined oil supplies has become practically as fundamental to social wellbeing as water and food, since supply and consumption of the latter two human staples is in large part dependant on the production and price of the former. For developing and developed economies alike, large increases in oil prices can have a major impact on the stability of states. Permanently high-energy prices, depending on the circumstances of a given society, may seriously undermine the socio-economic security of its people and, as a consequence, the security of the government and the state itself. The effects of such transnational phenomena on the internal circumstances of developing states in particular have the potential to fundamentally alter both the nature and stability of the state, which inevitably will be reflected in its policies and relationships with other states.

In this chapter, I use international energy price increases and shortages as an example of a transnational threat to argue that a) such threats should be recognised as threats to both individuals and the state; and b) doing so provides opportunities for better understanding transnational threats by using the internal and human need factors emphasised by human security approaches to establish a broader but still analytically useful security paradigm. Such a synthesis, however, requires a clearer definition of human security if it is to make any significant contribution to our conceptualisation of how, and the extent to which, the security of both the state and its people is threatened by transnational challenges like sharp energy price increases. Paramount among the criticisms made of the human security approach, with its relatively new focus on the 'security' of groups and individuals as opposed to the state, is the perception that 'human security' is too vague and all encompassing a concept to be of analytical use, despite its potential to address the kinds of issues mainstream state-centric paradigms have long ignored. As one critic has noted, 'if human security means almost anything, then it effectively means nothing' (Paris, 2001: 93).

Thus, what is needed to bridge the divide between the state and its people, in terms of how national security is perceived and pursued, is a clearer and more carefully defined conceptualisation of 'human security' and its capacity to contribute to the equally important, but increasingly limited, perspectives provided by traditional international relations approaches.[1] My analysis presents a socio-economic-based[2] human security definition – emphasising the importance of state facilitation and maintenance of socio-economic rights as a major determinant of both government and state legitimacy – as a possible foundation for the further development of a joint human security/state security paradigm able to identify and manage a wider range of potential threats and their effects. What follows then is an assessment of the problems and threats an energy crisis could create for states at various stages of socio-economic and political development in the Asia–Pacific that has been assembled using a combination of human security, human development, and international relations perspectives. These ideas are offered up as a departure point for further research and discussion on the need to better conceptualise and define 'human security', and also more clearly understand the nature and potential consequences of transnational threats, particularly in regard to oil price increases.

Transnational threats and state legitimacy: bridging the human security-state security divide

Transnational threats to the security of individuals and states are certainly not new; one only need think of the human devastation caused by the 1919 influenza pandemic or the global economic chaos caused by the 1929 Wall Street crash to illustrate the continued existence of cross-border threats that develop independently of human design. Over recent years there has been a growing awareness, encouraged in part by the effects of economic interdependency, of how the destabilising effects of non-state generated security challenges or internal crises in one state or region can easily spill over borders and jurisdictions, compromising the internal stability of other societies in ways that national governments usually have no direct means of addressing or controlling. The Asian financial crisis in 1997–1998 is one of many recent examples of how the security of individuals, groups, and the state itself remains vulnerable to forces that can develop unnoticed by state-based threat perceptions. Other contemporary examples include global climate change, cross-border pollution, exploitation of marine resources, in particular fisheries, and transnational terrorism.

What is new then is the much closer attention now being paid to these phenomena, as illustrated by the current preoccupation of many governments, academics, and the media with managing and preventing pandemics and terrorism, which has been provoked by recognition among some policy-makers and analysts of the need for transnational issues to be understood as distinctly different from state-centric forms of threats. Indeed, it was not until the early post-cold war years that the level of public and scholarly attention attracted by transnational issues, beyond their immediate impacts, began to rival that attracted by wars, ideological conflicts, and international trade and commerce. But this is not surprising given that our perceptions of security long have been dominated by an unyielding focus on the 'state' and the business of defending its territorial integrity and sovereignty from the expansionist inclinations of other states. The strong influence of such state-centred thinking, itself a product of the rational-choice-based tradition of mainstream international relations theory, has kept transnational and human security concerns outside the realm of mainstream 'national security' threats, which still remain largely limited to the strategic intentions, both militarily and to a lesser degree economically, of other government actors.

My intention is not to discount or downplay traditional threat perceptions, but to argue that because externally focused, state-based threat perception models – such as those provided by realism – are inadequate when dealing with transnational issues and the dangers they pose, we need to expand our understanding of 'national security' to include non-conventional and non-state generated threats. By defining human security as the socio-economic security of individuals, and recognising that transnational threats undermine the security of the state only because of the direct impact they have on human security, the 'bottom up' process through which the stability and security of the state and its people are intertwined and fundamentally inseparable becomes clear. Moreover, doing so establishes the two-way relationship between the security and welfare of individuals on the one hand and the national security of the state on the other. State failures to protect their citizens from human security challenges, as I define them,[3] affect not only the welfare of individuals but also the legitimacy and stability of the state.

Thus, it is a mistake, I contend, to view traditional notions of security and the emerging focus on human security as incompatible, or to simply dismiss all potential 'human security' perspectives as conceptually flawed and therefore inappropriate for research and analysis. And although the post-Cold War world has introduced many new, and some not so new, challenges to the traditional concept of the state and its usefulness in understanding the current nature of international relations, it would also be a mistake to view current state-centric threat perceptions as redundant. The rational theory-based approach underpinning neo-liberal and neo-realist theories needs instead to be expanded by taking on board the essential human security argument that it is people, rather than the abstract of the state, that represent the ultimate referent for security. The result would be a far more comprehensive concept of security – one that treats the state as a means to an end rather than an end in itself.

Conceptualising the human security-state security interface: legitimacy and socio-economic security

The currently broad concept of human security, which has evolved out of the United Nations Development Programme's flagging in 1994 (UNDP, 1994) of the need for a more people-orientated approach to security, owes much to the ideas and debates within human development discourse. And although the human development literature certainly does not ignore the importance of first generation political freedoms, second generation socio-economic rights are what most of the human development solutions on offer are focusing on. The Millennium Development Goals (UNDP, 2000) for example, are themselves almost entirely concerned with the improvement of socio-economic factors such as health, education, poverty and unemployment, environmental degradation, trade, and financial and economic reform.

The idea of human security, as it has evolved and is currently being pursued at the international level at least, has always prioritised the provision of basic needs over political freedoms.[4] It is on this basis, therefore, that definitions of human security should remain limited to issues directly affecting the socio-economic security of individuals and groups. Defining human security in this way makes it consistent with the widely recognised importance of socio–economic factors for human development initiatives, and also compliments the strong link I make between the presence or absence of socio-economic fundamentals and levels of state legitimacy and stability. But this linkage needs to be located within the context of 'national security' in order for the relationship between the socio-economic security of individuals and the stability of the state to become clear.

Governments committed to the realisation of stable, prosperous economies and satisfactory levels of socio-economic security are seldom confronted with serious internal challenges[5] to their authority or legitimacy only on the basis of concerns over civil and political rights. Challenges to government legitimacy that include demands for the kinds of political and civil human rights outlined in international treaties, which are themselves based largely on liberal democratic values and ideals, are usually symptomatic of government failures to provide expected levels of socio-economic security, rather than simply any innate and universal desire for Western-defined political freedoms. As Gray argues, challenges to state legitimacy are determined less by the system of government – and whether it is democratic, monarchical, or authoritarian – than they are by the record of that particular government or system of government in meeting socio-economic security expectations:

> Governments are legitimate in so far as they meet the needs of their citizens. Those that fail in this will be judged by their citizens to be illegitimate

whether or not they are democracies. People everywhere demand from governments security against the worst evils: war and civil disorder, criminal violence, and lack of the means of decent subsistence. How a state performs this protective role is the core test of its legitimacy. Unless it is discharged competently no other criterion can come into play. Thus, it is not whether a state is a liberal democracy that most fundamentally determines its legitimacy; it is how well it secures its citizens against the worst evils.

(Gray, 1998: 149–150)[6]

Thus, democracy, and the various negative rights that its numerous manifestations may or may not entail,[7] is a potential product of successfully implemented human development and security initiatives, rather than a catalyst for them; Przeworski and Limongi's (1997) empirical argument that development does not necessarily lead to democracy, but only makes democracies less likely to fail once they have been established appears to support the important relationship between socio-economic development and political stability.[8] Interestingly, in the context of this analysis, their conclusions also may suggest that democracies established at early or transitional stages of development are more vulnerable to destabilisation by transnational threats than those developed at a later stage of development, which raises a number of issues concerning the timing of democratic transitions, in addition to the highly controversial question of whether or not democratic systems of government in developing societies are more or less resilient to socioeconomic shocks, vis-à-vis challenges to legitimacy, than more authoritarian systems such as China's.[9]

But regardless of the long-running debate over the desirability and universality of democratic principles, their importance in terms of internal state legitimacy is much less than the state's ability and willingness to provide socio-economic security to its people. Developed and post-industrial societies have, to a greater or lesser degree, established democratic principles and practices that are reinforced and further developed through the maintenance of human security for at least the majority. The provision of human security, therefore, plays a critical role in supporting both the stability of the state and its system of government, regardless of its flavour. Indeed, when looked at in this way, the importance of socioeconomic security to 'state security' becomes clear: the stability of the state is heavily influenced by the living conditions and sense of security of its people. In the context of human security and the important role it plays in providing internal stability to states, the kind of government in power, as opposed to what it provides, is of little significance.

And while success or failure in this regard directly affects a state's internal stability, it is also important to note that the ramifications for the state do not end there. Severe socio-economic crises within a state also undermine its power, influence, and independence externally as an international actor. As Kim has argued, using North Korea as a case in point, significant levels of internal socioeconomic insecurity also directly undermine state security at the international level by making states more dependent on other states and international actors and severely limiting foreign policy choices and actions (Kim, 2001: 33–34). The fall of the Soviet Union's authoritarian government and the ensuing break-up of the former Soviet state is perhaps the most compelling example of how a deterioration of internal social and economic circumstances can set off a legitimacy crisis capable of completely destroying the existing state structure. Such examples demonstrate how both the internal, non-state orientation of human security can be adopted to cover the many blind spots that undermine the utility of existing notions of national security, and also the very great extent to which the internal dynamics produced by internal human security problems can represent a very real threat to not only the stability of the state but potentially also its existence.

The human security dimension of transnational threats

Because transnational threats generally are not state based and do not involve the use of violence or coercion by one state against another, they mostly exist outside of conventional notions of 'national security', or are awkwardly accommodated within them, as has been the case with transnational terrorism. Policy-makers and analysts have either resisted the inclusion of the many potential transnational threats under the national security banner, or they have selectively included some but not others as illustrated by the priority now given to terrorist threats in the national security strategy of the United States.

It is, however, important to note two points concerning the high-level policy attention that transnational terrorism attracts as a recognised national security threat. Firstly, the priority given to transnational terrorism does not stem from its status as a transnational threat, but rather from the *level of threat it poses* following the 9/11 attacks on the US, which is consistent with my approach to defining human security threats, transnational or otherwise, on the basis of their potential to disrupt the socio–economic security of individuals rather than any taxonomy-based criteria. Secondly, transnational terrorism's inclusion as a threat to national security under state-based notions of what constitutes national security has, in terms of policy responses, resulted in it being conceptualised and dealt with in the same way as conventional national security threats rather than as the entirely different form of threat it constitutes.

Terrorist threats and attacks are, as a result of this inflexibility, being treated as though they are a) state based in their origin (e.g. the War on Terror, itself a product of traditional security perceptions, has mostly involved the use of conventional military force against states)[10]; and b) a direct threat to the state (i.e. the US), as opposed to a direct threat on the human security of the American people, which is what generates the actual threat to the state vis-à-vis domestic perceptions of the legitimacy of US foreign policy and government responses to the threat. Many transnational threats, therefore, are either overlooked or misrepresented within existing security paradigms because they are neither state based nor are their impacts necessarily state focused. They are, however, essentially human security threats with the potential to affect state security because they directly affect the socio-economic security of individuals – not the state as a whole –

thereby constituting an indirect threat to the state through their impact on perceptions of state legitimacy.

Thus, in the context of transnational threats, the socio–economic security of people – as opposed to the economic and political security of the state and its government – is the first point of impact.[11] That is to say, the potentially negative impacts of transnational phenomena that occur independently of any state policy specifically intended to produce such impacts – what I define as 'transnational threats' – *directly* affect only the kinds of security issues that fall within the category of so-called positive human rights, which by definition require action or intervention by governments to be realised or protected. This approach, as indicated earlier, provides an essentially 'bottom up' perspective, which identifies the welfare of people as the key to the security of the state, rather than the 'top down' focus of state-centric security analysis, which only recognises threats in terms of their impact on the state.

In addition to the ways transnational disturbances can impair a state's ability to provide or protect the socio-economic needs of its people, the extent to which it is able to do so in reacting to a threat will ultimately determine the legitimacy of the government, and potentially the state itself. Human security challenges, such as transnational threats, therefore, are also state threats – depending on the level of impact they are likely to have on the socio–economic security of individuals within the state – and need to be included within a much more rigorous and sophisticated conceptualisation of 'national security' than currently exists. On the basis of the approach I have outlined here, the level of socio–economic (read human) development achieved by a state is the critical factor in determining the extent to which various transnational phenomena can threaten its stability and integrity in much the same way that the military and economic power of a state determines its vulnerability to conventional state-based threats.

Oil price impacts on human and state security: four illustrations of development-related vulnerability

The ability of governments to manage non-state-based threats to human security, such as natural disasters, pandemics, and financial and energy shocks, and prevent them from destabilising the state of course depends on the severity and nature of the crisis. But, based on the analysis I have presented here, the ability to absorb transnational and other non-state-based impacts like sharp increases in energy costs is also determined by the stage of development a state is in and the level of socio-economic security expected by its people. Thus, the states at most risk to human security threats are, not surprisingly, those that are either relatively undeveloped or developing, since their ability to react to crises is considerably less than that of the richer, and better organised developed states. Energy-based threats like the current wave of escalating oil prices, however, create a situation where the least exposed societies are both developed and undeveloped states, since the former have decreasing oil reliance, alternative fuels and technology, and greater capacity to absorb higher costs and the latter have little reliance on oil because they are not yet industrialised. The states most vulnerable to energy shocks, therefore, are the in-between, aspirational states such as China and Indonesia.

In Asia the legitimacy of both the Chinese Communist Party (CCP) and the Indonesian government, for example, are heavily dependent on their success in modernising and developing their respective societies, which in other words equates to progress in improving human security. The CCP is pursuing socio-economic development in a situation where it has constructed a national identity legitimised by modernisation. Indeed, questions of national identity and government legitimacy in both China and Indonesia are now primarily about the creation of an economically modern and prosperous society that is able to guarantee the security of its citizens; survival for both governments essentially depends on how good a job their respective populations believe they are doing in realising that goal. But as the fall of Suharto after the Asian financial crisis demonstrated, governments can quickly become endangered by sharp reductions in human security. The Suharto regime's fate indicates that even authoritarian control can be ineffective when the gap between human security expectations and actual living standards reaches a particular point, especially in a state like Indonesia where endemic government corruption and nepotism were seen to be adding to the hardship of people affected by the crisis.

In order to illustrate the argument I have made concerning the importance of human security levels and expectations in determining the impact of transnational threats, I have chosen four states in the Asia–Pacific that are representative of the three generally recognised stages of economic and social development: developed/post-industrial – Japan; aspirational/developing – China and Indonesia; and (relatively) undeveloped – Samoa. I am using these examples to demonstrate my contention that a state's ability to establish and maintain acceptable levels of human security determines the extent to which transnational threats, in this case energy shortages and price increases, can directly undermine the security of the state internally (in terms of its legitimacy and internal cohesion) and also lead to greater pressure for the state to then adopt external policies that may bring it into conflict with other states.

The economic dangers of sharp energy price increases are well known, as are the threats they pose to the security of states and individual citizens alike. Inflation, higher unemployment, sharply reduced economic growth, and reduced fiscal flexibility are all recognised outcomes of energy price increases that have obvious implications for the socio-economic standards of societies. Of the four states chosen, Japan is clearly the least vulnerable due to its ability to absorb oil price increases with relatively little impact on its human security levels. As a post-industrial society, Japan's oil dependence, in contrast to China and Indonesia, is decreasing. Japan also enjoys the benefit of alternative energy sources and the ability to easily switch between them. Japanese infrastructure is also highly developed and efficient, particularly in terms of public transport and telecommunications, which is a major advantage in terms of being able to reallocate and conserve oil supplies within essential sectors of the economy without any major disruption to the daily lives and routines of most people. As a long-established, stable democracy based upon a strong sense of national identity, any negative effects and hardships caused by energy shortages will be unlikely to pose any major threat to the stability of Japanese society and the state.

As aspirational states, still in the middle ground between development and non-development, China and Indonesia are the most exposed to the threats to human security posed by rising energy prices and supply shortages. In contrast to Japan, China and Indonesia's energy needs are increasing and neither country has the advantage of being able to quickly switch between energy sources or to fall back on any sizable alternative domestic sources such as nuclear or hydro-electric power. Furthermore, both countries are far less efficient in their energy usage, and lack the same level of highly sophisticated infrastructure (e.g. extensive rail electrification and telecommunications) that provide Japan with alternatives in keeping the economy and essential services operating in spite of oil shortages. Of particular significance, especially in the case of Indonesia, is the prevalence of government-funded oil subsidies in developing economies.

Unlike developed economies, which tax rather than fund energy consumption, developing economies are caught in a catch-22 situation where either continuing or eliminating fuel/energy subsidies will undermine human security within society. This double bind results from a dilemma between human security issues on the one hand and fiscal responsibility on the other. If Indonesia's government makes further cuts to its fuel subsidies, more fully exposing Indonesians to international oil prices, it will reduce living standards for many of its people – and risk an escalation of the already widespread protests against existing subsidy cuts – by setting in play the kinds of economic effects mentioned above. The upside is that people will become more efficient in their use of energy, but for many low-income groups their already financially insecure situations may deteriorate into destitution and poverty, as occurred in the aftermath of the 1997–1998 financial crisis. Alternatively, if the government attempts to continue subsidising energy prices, it will, in addition to losing any incentive for more efficient energy use, risk fiscal ruin and result in other areas of government support, such as health and education, also being adversely affected.

The Chinese government has lessened its exposure to the subsidy dilemma by slowly reducing its levels of subsidisation over the last 10–15 years, but its rapid economic development has created a domestic level of demand for energy that is now outstripping the level of supply available to China. China and Indonesia's aspirational status means that ongoing economic development is of crucial importance, politically and socially, to both governments. The legitimacy of each government is tied to their success in ensuring continuing development, and failure on this score will risk major dissent and instability within their respective societies. The problems posed by higher energy prices and shortages represent a clear threat to the developmental process, and pressures policy-makers to maintain a level of energy supply that will not endanger further development. Thus, internal imperatives for development and human security begin to exert pressure on the state that then extends into the external realm of inter-state relations. This internally driven influence on foreign policy is already being reflected in China's relentless pursuit of foreign-based energy supplies and its increasingly belligerent relationship with Japan over access to oil and gas resources in the East China Sea, and could force Indonesia into a similar approach over resources in the South China Sea. Indonesia's recent transition to an 'aspirational' democracy, as opposed to China's ongoing authoritarian system of government, also leaves the state more exposed to the threat of internal instability due to the many competing notions of national identity that a more open political society allows for, particularly because of the country's great cultural and ethnic diversity. Tolerance for human security impacts in Indonesia, therefore, is likely to be lower than in China where the Communist Party still has control over notions of national identity and also a greater ability to stifle dissent.

Like Japan, Samoa too will be better placed than either China or Indonesia to absorb energy shortages, albeit for entirely different reasons. As a relatively undeveloped society, Samoa's oil dependence is low (nearly half of Samoa's electricity needs are provided by hydro-electric plants) and, therefore, oil price increases and shortages have very limited potential for significantly reducing already existing levels of human security. Moreover, unlike in China and Indonesia, modernisation and development expectations in Samoa are also low, and are of far less political significance in terms of government legitimacy and national identity issues. As a country heavily reliant on foreign aid, however, Samoan human security is vulnerable to indirect fallout of an energy shock that could result from aid contributions reducing or even stopping as a result of economic problems brought on by energy shortages in the donor countries. That said, Samoa's human security and the potential for its decline to destabilise the state as the result of an energy shortage is of a much lower order of probability than that facing Indonesia and China.

Conclusion

Like many other forms of transnational threat, energy shortages and price increases have a direct impact on human security in all societies. These impacts have the potential to lead to both the destabilisation of the state internally and greater pressure on governments to adopt foreign policy stances that are seen as provocative by other states and/or increase its dependency on other states and international assistance. But the extent to which a deterioration of human security levels can destabilise the state is determined by whether it is in a transitory stage of economic development and possibly also by whether its economic and social transition is being accompanied by political transition.

In terms of promoting and developing human security, as per the socio–economic factors I have used to define it here, economic development is essential for not only establishing acceptable levels of security, but also for providing the means for the state to protect the security of its citizens when threats arise. Development, or non-development in the case of energy, therefore, is a basic requirement for managing and reacting to non-state threats that target individuals

Socio-economic, geographic, and energy consumption/production comparison data

	Japan	China	Indonesia	Samoa
Land area (sq km)	374,744	9,326,410	1,826,440	2,944
Population (July 2005 est.)	127,417,244	1,306,313,812	241,973,879	177,287
Infant mortality rate (2005 est.) (deaths/1,000 live births)	3.26	24.18	35.6	27.71
Literacy (age 15 and over can read and write, total population)	99% (2002)	90.9% (2002)	87.9% (2002)	99.7% (2003 est.)
GDP per capita PPP	$29,400 (2004 est.)	$5,600 (2004 est.)	$3,500 (2004 est.)	$5,600 (2002 est.)
Unemployment rate (2004 est.)	4.7%	20% total 9.8% in urban aeas	9.2% (2004 est.)	n.a. (sustainable underemployment)
Population below poverty line	n.a.	10% (2001 est.)	27% (1999)	n.a.
Oil production (bbl/day)	17,330 (2001 est.)	3,392 million (2003 est.)	971,000 (2003 est.)	0
Oil consumption (bbl/day)	5.29 million (2001 est.)	4.956 million (2002 est.)	1.183 million (2003)	1,000 (2001 est.)
Natural gas production (billion cubic metres)	2.519 (2001 est.)	35 (2003 est.)	77.6 (2003 est.)	–
Natural gas consumption (billion cubic metres)	80.42 (2001 est.)	29.18 (2002 est.)	55.3 (2003 est.)	–
Electricity production by source (2001)				
• fossil fuel	60%	80.2%	86.9%	58%
• hydro	8.4%	18.5%	10.5%	42%
• nuclear	29.8%	1.2%	0%	0%
• other	1.8%	0.1%	2.6%	0%
Telephone – main lines	71.149 million (2002)	263 million (2003)	7.75 million (2002)	11,800 (2002)
Telephone – mobile cellular	86,658,600 (2003)	269 million (2003)	11.7 million (2002)	2,700 (2002)
Internet users	57.2 million (2002)	94 million (2004)	8 million (2002)	4,000 (2002)
Railways total	23,577 km (2004)	71,898 km (2002)	6,458 km (2004)	–
Railways electrified	16,519 km (2004)	18,115 km (2002)	125 km (2004)	–
Highways total	1,171,647 km (2001)	1,765.222 km (2002 est.)	6,458 km (1999)	790 km (1999 est.)
Highways paved	903,340 km* (2001)	395,410 km* (2002)	158,670 km (1999 est.)	332 km (1999 est.)

Notes: *Including 6,851 km of expressways in Japan and 25,130 km of expressways in China.

Source: The World Factbook http://www.cia.gov.cia.publications/factbook. Accessed 17 August 2005.

and groups rather than the state itself, but the ability to maintain human security needs to be accompanied by a broader conceptualisation of 'security' that recognises the interdependence of state *and* individual security. Traditional state-centric perceptions of security are limited to state-based threats and responses, leaving these approaches unable, for the most part, to conceive of, let alone effectively manage, transnational threats that, by definition, often cannot be traced back to a particular or tangible source.

In the context of energy shortages and price increases, transitional economies are most at risk because of their high energy dependence and lack of flexibility and infrastructure, on the one hand, and, on the other, the political vulnerability that comes from being in a position where the legitimacy of the state, in the eyes of its people, is dependent on its ability to provide the socio-economic improvements that underpin people's human security expectations. This places aspirational states like China and Indonesia on something of a tightrope in that the tolerance in their societies of even relatively small transnational disturbances is considerably less than that of either developed or undeveloped societies.

Indonesia, on the basis of my analysis, faces the greatest risk of economic and political destabilisation from energy price increases due to its high levels of energy subsidies and the fiscal problems this practice now presents. Governments in the region with a strong interest in ensuring the continuing development in Indonesia of both its economy and fledgling democratic system of governance may, therefore, need to consider providing economic aid to Indonesia to allow the government's current efforts at energy subsidy reductions to continue without seriously eroding the socio–economic security of the Indonesian people. Indeed, the importance of carefully gauging the potential effects of further oil price increases and the Indonesian government's ability to provide a 'soft landing' for its people should not be neglected if the kind of political backlash that led to the demise of the Suharto government is to be avoided.

Notes

1 For a good overview and treatment of discussions on the possible melding of human and traditional security approaches, see Kim (2001).
2 Used here in reference to the generally accepted social goals, included in the 1966 *International Covenant on Economic, Social, and Cultural Rights*, consisting of rights to food, health, shelter, education, work, and protection from social/environmental harms such as illegal drug trafficking, violence, disease, and social discrimination or persecution on the basis of race/culture/sex.
3 I am defining human security challenges/threats only in terms of their (potential) negative effects on the socio-economic expectations of individuals.
4 The UNDP's *Human Development Report 1994* clearly locates its concept of 'human security' within a human development paradigm that 'puts people at the centre of development', emphasising that 'In the final analysis, sustainable human development is pro-people, pro-jobs and pro-nature. It gives the highest priority to poverty reduction, productive employment, social integration and environmental regeneration' (UNDP, 1994: 4).
5 This of course does not exclude challenges from other states over alleged breaches of negative human rights responsibilities.

6 In a similar vein, Jack Donnelly (1999: 609) asserts that:
   The link between a regime's ability to foster development (prosperity) and the public's perception of the regime's legitimacy is close to a universal, cross-cultural law. Whatever a ruling regime's sociological and ideological bases, its sustained or severe inability to deliver prosperity, however that may be understood locally, typically leads to serious political challenge.

7 The democratic transition paradigm and the standards it uses for assessing progress towards democracy approach has recently been questioned and heavily criticised by Thomas Carothers and others on the grounds that its conception of 'transitional' democracies is ill-defined and exaggerates the benefits of establishing political processes (e.g. elections, political parties) while dismissing the importance of underlying conditions and structural features such as levels of economic and institutional development, political history, and socio-cultural traditions (Carothers, 2002).

8 See Przeworski and Limongi (1997); see also Boix and Stokes (2003) for a discussion of Przeworski and Limongi's conclusions and methods. Boix and Stokes reject Przeworski and Limongi's claim that democratic transition is an exogenous rather than endogenous process, arguing instead that socio-economic development is an essential ingredient for endogenous democratic transition.

9 The major factor affecting government and state legitimacy, according to the argument I am presenting, remains the level of socio-economic security provided by the state. In this context, the political persuasion of the government is not significant, since legitimacy is tied to socio-economic rather than ideological criteria. In the context of transnational threats, the issue of preferring one system of government over another is only relevant in so far that it influences human development outcomes.

10 In addition to the fact that most of the money spent on prosecuting the WOT has gone into military operations in Iraq and Afghanistan, George W. Bush made the state-based focus of the WOT very clear from the outset with his 'with us or against us' declaration in early November 2001.

11 For example, al Qaida, pollution, climate change, pandemics, and natural disasters. Conversely, the consequences of transnational impacts on negative human rights (i.e. those requiring an absence of intervention by governments), such as political and civil rights, are either of a lower priority (such as in disaster relief efforts) or occur only indirectly as a result of government measures introduced to deal with the threat posed to socio-economic rights (such as, for example, the strengthening of anti-terrorism laws or declarations of martial law).

Bibliography

ABARE (Australian Bureau of Agricultural and Resource Economics), 2005, *Energy Security in APEC*, ABARE Publications, Canberra.

Abelson, A., 2005, Friendly Dragon? *Barron's*, 27 June.

Abhyankar, J., 2001, 'Piracy and Shipping Robbery: A Growing Menace', in H. Ahmad & A. Ogawa (ed.) *Combating Piracy*, The Okazaki Institute, Tokyo.

Abraham, S., 2004, Speech entitled 'US National Energy Policy and Global Energy Security', *Economic Perspectives* accessed on August 23 2005, available at: http://U.S.info.state.gov/journals/ites/0504/ijee/abraham.htm.

'Action Not Word', 2004, *Power line*, March.

ADB (Asian Development Bank), 2004, *Asian Development Outlook 2004 Update*, ADB, Manila available at: http://www.adb.org/Documents/Books/ADO/2004/update/part030200.asp

Adelman, M. A., 2004, 'The Real Oil Problem', *Regulation*, Spring.

Alford, P., 2005, 'Oil Will Flow, Putin Tells Japan', *The Australian*, 22 November.

Alhajji, A. F., 2005, *Journal of Energy and Development*, 30 (2), 223–237.

'Ali al-Naimi calls for building more refineries in consumer countries', 2005, *Alexander's Gas and Oil Connections*, 10 (3), 6 July.

al-Yabhouni, A., 2005, 'OPEC Fulfilling its Commitments but Refining and Shipping Bottlenecks Keep Oil Prices on Boil', *Middle East Economic Survey*, xlviii (19).

Andrews-Speed, P., 2004, *Energy Policy and Regulation in the People's Republic of China*, Kluwer Law International, The Hague.

——, 2005, 'China's Energy Woes: Running on Empty', *Far Eastern Economic Review*, June.

Andrews-Speed, P., Liao X. L., & Dannreuther, R., 2004, 'Searching for Energy Security: The Political Ramifications of China's International Energy Policy', *China Environment Series*, 5, Woodrow Wilson International Centre for Scholars, January.

APCSS (Asia–Pacific Center for Security Studies), 1999, 'Energy Security in the Asia–Pacific: Competition or Cooperation?' Executive Summary, January 15, available at: http://www.apcss.org/Publications/Report_Energy_Security_99.html.

APEC Energy Research Centre, 2003, *Energy Security Initiative: Some Aspects of Oil Security*, Institute of Energy Economics, Tokyo.

APERC (Asia–Pacific Energy Research Center), 2002, *Energy Security Initiative: Emergency Oil Stocks as an Option to Respond to Oil Supply Disruptions*, Asia–Pacific Energy Research Center, Tokyo, available at: http://www.ieej.or.jp/aperc/2002pdf/OilStocks2002.pdf

APPCDC (Asia–Pacific Partnership on Clean Development and Climate), 2005, Joint Press Release, *Australia Joins New Asia–Pacific Partnership on Clear Development and Climate,* available at: http://www.pm.gov.au/news/media_releases/media_Release 1482.html

Arndt, H. W., 1974, 'Resources Diplomacy', *Research Paper*, Australia–Japan Economic Relations Research Project, Australian National University, Canberra, July.

ARNE (Agency for Natural Resources and Energy), Ministry of Economy, Trade and Industry, 2005, *Energy in Japan 2005*, Communications Office, Agency for Natural Resources and Energy, Ministry of Economy, Trade and Industry, Tokyo.

Arvanitopoulos, C., 1998, 'The Geopolitics of Oil in Central Asia', *Thesis*, Winter.

Auliciems, A., & Burton, I., 1973, 'Trends in Smoke Concentrations Before and After the Clean Air Act of 1956', *Atmospheric Environment*, 7, 1063–1070.

Austin, A., 2005, *Energy and Power in China: Domestic Regulation and Foreign Policy*, The Foreign Policy Centre, London.

Australian Government, 2004, *Securing Australia's Energy Future*, available at: www.pmc.gov.au/energy_future

Bahgat, G., 2003, 'The New Geopolitics of Oil: The United States, Saudi Arabia, and Russia', *Orbis*, 47 (3), 447.

——, 2004, 'Russia's oil potential: prospects and implications', *OPEC Review,* June, 29 (2), 133–147.

——, 2005, 'Energy Partnership: China and the Gulf States', *OPEC Review*, June, 115–131.

Bajpaee, C., 2005a, 'Setting the Stage for a New Cold War: China's Quest for Energy Security', 25 February, accessed on 4 March 2005, available at: www.pinr.com

——, 2005b, 'China Fuels Energy Cold War', *Asia Times*, 2 March.

——, 2005c, 'China's Quest for Energy Security', ISN Security Watch, 25 February, available at: http://www.isn.ch

Barclay, Glen St. J., 1978, 'Australia and the new international economic order: a prospect of resources diplomacy', *World Review*, 17.

Barnett, D. W., 1979, *Minerals and Energy in Australia*, Cassell Australia, Stanmore, NSW.

Barraclough, G., 1974, 'The End of an Era', *New York Review of Books,* 21(11), 27 June.

Barsky, R., & Kilian, L., 2004, 'Oil and the Macro-economy since the 1970s', *Journal of Economic Perspectives,* 18 (4).

Barton, B., Redgwell, C., Ronne, A., & Zillman, D.N., 2004, *Energy Security: Managing Risk in a Dynamic Legal and Regulatory Environment*, Oxford University Press, London.

Belgrave, R., 1985, 'The Uncertainty of Energy Supply in a Geopolitical Perspective', *International Affairs*, 61 (2), 253–261.

Berlin, D. L., 2004, 'The "great base race" in the Indian Ocean littoral: conflict prevention or stimulation?' *Contemporary South Asia*, 13 (3), 239–255.

Bi, J., 2005, 'Strategic reserve to see oil by year's end', *China Daily*, 5 July, available at: http://www2.chinadaily.com.cn/english/doc/2005-7/05/content_457194.htm

Blagov, S., 2004, 'Russia Tangles with Japan and China', *Asian Times*, September 1.

——, 2005, 'Russia walks thin line between Japan and China', *Asia Times Online*, accessed on 27 August, available at: http://www.atimes.com/atimes/Central_Asia/GA05Ag01.html

Blank, S., 2004, 'India and the Gulf After Saddam', *Strategic Insights*, 3 (4).

——, 2005, 'India's Quest for Central Asian Energy', *The Jamestown Foundation Eurasia Daily Monitor*, 40 (2).

Boeicho Boei Kenkyusho (ed.), 2005, *Higashi Ajia Senryaku Gaikan* (East Asian Strategic Review 2005), Boei Kenkyusho, Tokyo.

Boix, C., & Stokes, S. C., 2003, 'Endogenous Democratisation', *World Politics*, 55, July.

Boussena, S., & Locatelli, C., 2005, 'Towards a more coherent oil policy in Russia?', *OPEC Review*, June, 29 (2), 85–105.

——, 2005, 'The bases of a new organisation of the Russian oil sector: between private and State ownership', accessed April, available at: http://web.upmf-grenoble.fr/iepe/textes/SB-CL_bases_JlofEnergyandDev05.pdf

BP, 2005, *BP Statistical Review of World Energy*, accessed June, available at: http://www.bp.com.

Bradshaw, M., 2003, 'Prospects for oil and gas exports to Northeast Asia from Siberia and the Russian Far East, with a particular focus on Sakhalin', *Sibirica: Journal of Siberian Studies*, 3 (1), April.

Brodman, J. R., 2005, US Department of Energy, Statement before the Subcommittee on International Economic Policy, Export and Trade Promotions, Committee on Foreign Relations, US Senate, October 27.

Bromley, S., 1994, *Rethinking Middle East Politics*, University of Texas Press, Austin.

——, 2005, 'The United States and the Control of World Oil', *Government and Opposition*, 40 (2), Spring.

Brzezinski, Z., 2004, *The Choice*, Basic Books.

Buchan, D., 2002, 'The Threat Within: Deregulation and Energy Security,' *Survival*, 44 (3), 105–115.

Byrnes, M., 1994, *Australia and the Asia Game*, Allen & Unwin, Sydney.

Calabrese, J., 2002, 'In the Shadow of Uncertainty: Japan's Energy Security and Foreign Policy', *Pacific and Asian Journal of Energy*, 12 (1).

——, 2004, 'Dragon by the Tail: China's Energy Quandary', *Middle East Perspective*, 23 March.

Calder, K. E., 1996, *Pacific Defense: Arms, Energy, and America's Future in Asia*, William Morrow and Company Inc, New York.

——, K. E., 1997a, *Asia's Deadly Triangle: How Arms, energy and Growth Threaten to Destabilize Asia–Pacific*, Nicholas Brealey Publishing, London.

——, 1997b, 'Fueling the rising sun', *Harvard International Review*, 19 (3), 24–30.

——, K. E., 2004, 'The Geopolitics of Energy in Northeast Asia', presented at the Korean Institute for Energy Economics, Seoul, accessed on March 16–17, available at: http://www.nautilus.org/fora/security/0432A_Calder.pdf

——, K. E., 2005, 'Toward a US–Japan Energy Policy Dialogue', *Gaiko Forum: Japanese Perspectives on Foreign Affairs*, 5 (1), 14–18.

Campbell, C., 2003, *The Coming Oil Crisis*, Multi-Science Publishing, Essex, Appendix 2.

Cannon, J. S., 2005, *The Transportation Boom in Asia: Crisis and Opportunity for the United States*, available at: www.informinc.org/rpt_asia.php

Carothers, T., 2002, 'The End of the Transition Paradigm', *Journal of Democracy*, 13 (1).

Caruso, G. F., & Doman, L. E., 2004, Energy Information Administration, US Department of Energy, 'Global Energy Supplies and the US Market', *Economic Perspectives*, accessed on August 23 2005, available at: http://U.S.info.state.gov/journals/ites/0504/ijee/carU.S.o.htm

Chang, F. K., 2001, 'Chinese Energy and Asian Security', *Orbits*, Spring.

Chanlett-Avery, E., 2005, *Rising Energy Competition and Energy Security in Northeast Asia: Issues for US Policy*, CRS Report for Congress February 9.

Chapman, J. W. M., Drifte, R., & Gow, I. T. M., 1982, *Japan's Quest for Comprehensive Security: Defence, Diplomacy, Dependence*, St Martin's Press, New York.

Chaturvedi, Sanjay (forthcoming), 'India's Quest for Strategic Space in the 'New' International Order! Locations, (Re)Orientations and Opportunities', in P. Jain, F.

Patrikeeff & G. Groot (eds.) *Asia–Pacific and a New International Order Responses And Options*, Nova Science, New York.

Chellaney, B., 2005, 'India, China mend fences', *The Washington Times*, April 7.

Chen Aizhu, 2005, 'China Fuel Tax its Best Weapon to Check Oil Demand', Reuters March 30, Climate Ark – Climate Change Portal, available at: http://www.climateark.org/

'China Hails India–ASEAN Summit', 2002, *Economic Times*, New Delhi, November 8.

Chow, D. C. K., 1997, 'An Analysis of the Political Economy of China's Enterprise Conglomerates,' *Law and Policy in International Business*, 28, 383–433.

Christoffersen, G., 2004, 'Angarsk as a Challenge for the East Asian Energy Community,' paper presented at 'Northeast Asian Security: Traditional and Untraditional Issues', Renmin University of China, 2–4 April.

Chung, C., 2004, 'The Shanghai Cooperation Organization: China's Changing Influence in Central Asia,' *The China Quarterly*, 990–1009.

Cohen, A., & O'Driscoll, Jr., G.P., 2002, 'The Road to Economic Prosperity for a Post-Saddam Iraq', *Heritage Foundation*, Backgrounder #1594, accessed on September 25, available at: http://www.heritage.org/Research/MiddleEast/bg1594.cfm

Cook, E., 1976, *Man, Energy, Society*, W.H. Freeman and Company, San Francisco.

Conant, M. A., & Gold, F. R., 1978, *The Geopolitics of Energy*, Westview, Boulder, Colorado.

Constantin, C., 2005, 'China's Conception of Energy Security: Sources and International Impacts', working paper No. 43 (March), Centre of International Relations, University of British Columbia.

Corden, W. M., 2005, Remarks at seminar on 'China, the US and the world economy', Australian National University (ANU), Canberra, 22 September.

Cossa, R., 2003, 'US Approaches to Multilateral Security and Economic Organizations in the Asia–Pacific' in R. Foot, N. S. MacFarlane & M. Mastanduno (eds.) *US Hegemony and International Organisations*, Oxford University Press (Oxford Scholarship on Line (OSO)), available at: http://www.oxfordscholarship.com/oso/public/content/politicalscience/0199261431/toc.html

Curtis, G. L., 1977, 'The Tyumen Oil Development Project and Japanese Foreign Policy Decision-Making', in R. A. Scalapino (ed.), *The Foreign Policy of Modern Japan*, University of California Press, Berkeley, pp. 147–173.

Daly, J. C K., 2004, 'The Dragon's Drive for Caspian Oil', *The Jamestown Foundation Eurasia Daily Monitor*, 4 (1).

Daniel, J., 2001, 'Hedging Government Oil Price Risk', *IMF Working Paper*, WP/01/185, November.

Dannreuther, R., 2003, 'Asian Security and China's Energy Needs', *International Relations of the Asia–Pacific*, 3.

Dawson, C., 2005, 'Nuclear Alert for Asia', *Far Eastern Economic Review*, October 14, 18.

Deese, D. A. & Nye, J. S. (eds.), 1981, *Energy and Security*, Ballinger Publishing Company, Cambridge, Mass.

Dempsey, J., 2004, 'Europe Worries Over Russian Gas Giant's Influence', *New York Times/International Herald Tribune,* accessed on October 4, available at: http://www.energybulletin.net/newswire.php?id=2389 and http://www.nytimes.com/2004/10/05/business/worldbusiness/05gazprom.html

Development Research Centre of the State Council, 2003, 'China's National Energy Strategy and Policy 2000–2020', accessed November, available at: http://www.efchina.org

Dienes, L., 2005, 'Observations on the Russian Heartland', *Eurasian Geography and Economics*, 46 (2), 156–163.

——, 2004, The Present Oil Boom (from his conference presentation); JRL #8399, accessed on October 7, Johnson's Russia List, available at: http://www.cdi.org/russia/johnson/8399-19.cfm

Dixit, A., 1992, 'Investment and Hysteresis', *Journal of Economic Perspectives*, 6 (1), 107–132.

DOE (United States Department of Energy), 2005, 'China Country Analysis Brief', accessed August, available at: http://www.eia.doe.gov/emeu/cabs/china.html

Donnelly, J., 1999, 'Human Rights, Democracy, and Development', *Human Rights Quarterly*, 21 (3).

Dorian, J. P., 2005, 'Growing Chinese Energy Demand: Dramatic Global Implications', Centre for Strategic and International Studies, Washington DC, 23 March.

Dorian, T. F., & Spector, L. S., 1981, 'Covert Trade and the International Nonproliferation Regime', *Journal of International Affairs*, 35(1).

Downer, A., 1996, 'Australia: Much More than a Middle Power', speech to the Young Liberals Convention, Canberra, 8 January.

——, 2003, 'The Myth of Little Australia', speech to The National Press Club, Canberra, 26 November, available at: http://www.foreignminister.gov.au/speeches/2003/031126_press_club.html

Downs, E. S., 2000, *China's Quest for Energy Security*, Rand Corporation, California.

——, 2004, 'The Chinese Energy Security Debate', *The China Quarterly*, 177.

Dowty, A., 2000, 'Japan and the Middle East: Signs of Change?', *Middle East Review of International Affairs*, 4 (4).

Drummond, M., 2005, 'Gas beats coal', *The West Australian*, 16 August.

Dupont, A., 2004, *Unsheathing the Samurai Sword: Japan's Changing Security Policy*, Lowy Institute for International Policy, Sydney.

Ebel, R., 2005, Is The Russian Oil Boom Over? Interview with Robert Ebel of CSIS, GlobalPublicMedia, Washington, 17 April, available at: http://www.globalpublicmedia.com/transcripts/382

Eckaus, R. S., 2004, 'Unravelling the Chinese Oil Puzzle', MIT CEEPR working paper, wp-2004-022, 29 December.

The Economist, 2005, 'Oil and gas in troubled waters', *The Economist*, accessed on 6 October, available at: http://www.economist.com/displaystory.cfm?story_id=E1_QQRJGVD

EIA (Energy Information Administration), 2004, *Top Petroleum Net Exporters, 2003*, US Department of Energy, Washington DC, August, available at: www.eia.doe.gov/emeu/security/topexp.html

——, 2005a, *International Energy Outlook 2005*, US Energy Information Administration, accessed July, available at: http://www.eia.doe.gov/oiaf/ieo/

——, 2005b, *Annual Energy Outlook 2005*, US Energy Information Administration, February.

——, 2005c, 'Country Profile: South Korea', available at: www.eia.doe.gov/emeu/cabs/skorea.html

——, 2005d, 'World Oil Markets', *International Energy Outlook (IEO)*, United States Government Printing Office, Washington DC.

——, 2005e, 'Country Analysis Brief: India', accessed on 23 June, available at: http://www.eia.doe.gov/emeu/cabs/india.html

——, 2005f, 'World Proved Reserves of Oil and Natural Gas, Most Recent Estimates', available at: http://www.eia.doe.gov/emeu/international/reserves.html

——, 2005g, *Russia Country Analysis Brief*, US Department of Energy, Washington DC, accessed February, available at: www.eia.doe.gov/emeu/cabs/russia.html

——, 2005h, 'Russian Oil and Gas Pipelines', available at: http://www.eia.doe.gov/emeu/cabs/russia_pipelines.html

Eklöf, S., 2005, 'Piracy in Southeast Asia: Real Menace or Red Herring?' *Japan Focus*, accessed 12 August, available at: www.japanfocus.org/article.asp?id=351

ElBaradei, M., 2004, Director General of the IAEA, 'Nuclear Proliferation and the Potential Threat of Nuclear Terrorism', *IAEA.org*, 8 November, available at: http://www.iaea.org/NewsCenter/Statements/2004/ebsp2004n013.html

'Energy Leading Group Set Up' 2005, *China Daily*, 4 June.

EUROGULF: An EU-GCC Dialogue for Energy Stability and Sustainability: Final Report, Kuwait, April 2005, Subtask 2.1.

Fackler, M., 2003, 'Japan: Hunt for Oil', *Far Eastern Economic Review*, March 20.

Fagan, R. H., 1981, 'Geographically uneven development: restructuring of the Australian aluminium industry', *Australian Geographical Studies*, 19 (2), 141–160.

Feigenbaum, E. A., 1999, 'China's Military Posture and the New Economic Geopolitics', *Survival*, 41 (2).

Feis, H., 1950, *The Road to Pearl Harbor: The Coming of the War Between the United States and Japan*, Princeton University Press, Princeton.

Feldman, S., 1982, 'The Bombing of Osiraq – Revisited', *International Security*, 7(2).

Fels, A., & Brenchley, F., 2005, 'No time to be over a barrel', *Australian Financial Review*, 5 May.

Fesharaki, F., 1991, 'Energy and the Asian Security Nexus', *Journal of International Affairs*, 53 (1), 85–99.

Finon, D., Locatelli C., & Mima, S. 2000, 'Long-term competition between gas infrastructure developments in Asia: The constraints on the Siberia and Caspian export development by cross-border pipelines', Institut D'Economie Et De Politique De L' Energie, Grenoble, accessed September, available at: http://web.upmf-grenoble.fr/iepe/textes/ dfclsm2000.pdf

Fitzgerald, B., 1992, 'ERA In Deal To Take Uranium From Kazakhstan', the Age, 31 October.

Foley, G., 1976, *The Energy Question*, Penguin Books, New York.

Foss, M. M., 2005, 'Global Natural Gas Issues and Challenges: A Commentary', *The Energy Journal*, 26 (2), 111–128.

Frankel, P., 1983, 'Oil: "Market Forces"', *Natural Resources Forum*, 7 (1).

Fu, K., & Li, B., 1995, 'Energy Development in China: National Policies and Regional Strategies', *Energy Policy*, 23 (2).

Gaddy, C. G., 2004, 'Perspectives on the Potential of Russian Oil', *Eurasian Geography and Economics*, 45 (5).

——, 2005, 'The End of Russia's Oil BoomæWhat Then?', *International Affairs* (Russia), 51 (1); Academic Research Library, p. 127.

Gaddy, C. G. & Hill, F., 2003, *The Siberian Curse: How Communist Planners Left Russia Out in the Cold*, Brookings.

'Gail Bid to Thwart Reliance Pipeline', 2002, *The Telegraph, Calcutta, India*, 5 August, accessed 7 July 2005, available at: http://www.telegraphindia.com/1040806/asp/business/story_3589398.asp

Garman, D., 2005, 'Testimony to the Senate Committee on Foreign Relations', July 26.

Geoscience Australia, 2005, 'Big New Oil – a progress report', *AusGeo News*, 77, available at: http://www.ga.gov.au/image_cache/GA6137.pdf

Giragosian, R., 2004, 'East Asian Tackles Energy Security', *Asia Times Online*, 24 August.

Giridhardas, A., 2005, 'India Proposes Transcontinental Pipeline to Chinese', *International Herald Tribune*, 19 April.

Goblet, Y. M., 1955, *Political Geography and the World Map*, F. A. Praeger, New York.

Goldman, M. I., 2004, 'Putin and the Oligarchs', *Foreign Affairs*, 83, (6), 33.

Goldstein, A., 2001, 'The Diplomatic Face of China's Grant Strategy: A Rising Power's Emerging Choice', *The China Quarterly*, 168.

Goodstein, L., 2005, 'Robertson Suggests US Kill Venezuela's Leader', *The New York Times*, August 24.

Gorst, I., 2003, 'Russia: Energy Strategy: Eastern Promise', *Petroleum Economist*, 70 (7), 10–11.

——, 2004, 'Kazakhstan oil to flow east', *Petroleum Economist*, London, December.

Gray, J., 1998, 'Global Utopias and Clashing Civilisations: Misunderstanding the Present', *International Affairs*, 74 (1).

Gries, P. H., 2005a, 'China Eyes the Hegemon', *Orbis*, 49 (3), 401–412.

——, 2005b, 'China's "New Thinking" on Japan', *The China Quarterly* (forthcoming).

Gwertz, B., 2005, 'China builds up strategic sea lanes', *The Washington Times*, January 17.

Hadley, S., 2006, US National Security Adviser, remarks to the National Bureau of Asian Research Strategic Asia Forum, The White House, 5 April.

Haglund, D. E. (ed.), 1989, *The New Geopolitics of Minerals*, University of British Columbia Press, Vancouver.

Haider, Z., 2005, 'Oil fuels Beijing's New Power Game', *Yale Global*, accessed on 11 March, available at: http://yaleglobal.yale.edu/display.article?id=5411

Hansen, J., Sato, M., Ruedy, R., Lacis, A., & Oinas V., 2000, 'Global warming in the twenty-first century: An alternative scenario', *Proc. Natl. Acad. Sci. USA*, 97 (18), 9875–9880.

Harris, M., 2003, 'Energy and Security', in M. E. Brown (ed.) *Grave New World: Security Challenges in the 21st Century*, Georgetown University Press, Washington, DC, pp. 157–177.

Harris, S., 2004, 'China in the Global Economy', in B. Buzan & R. Foot (eds.) *Does China Matter? A Reassessment*, Routledge, London and New York.

Harris, S., & Naughten, B., 2005, 'Economic Dimensions of Energy Security in the Asia–Pacific', paper presented to Workshop, *Energy Security in the Asia–Pacific Region* hosted by the Griffith Asia Institute and the Asia–Pacific Futures Network; Stamford Plaza Hotel, Brisbane, 31 August and 2 September.

Herberg, M. E., 2004, 'Asia's Energy Insecurity: Cooperation or Conflict?' in A. J. Tellis & M. Wills (eds.) *Confronting Terrorism In the Pursuit of Power*, National Bureau of Asian Research, Seattle.

——, 2005a, 'Asia's Energy Insecurity, China, and India: Implications for the US.', Testimony to the Senate Committee on Foreign Relations, July 26.

——, 2005b, 'Asia's Energy Insecurity', Presentation at National Bureau of Asian Research, APERC Energy Research Conference, Tokyo, Japan, accessed on 23 February, available at: http://www.nbr.org/programs/energy/APERC_Conference.pdf

Hill, F., 2002, 'Russia: The 21st Century's Energy Superpower?', *The Brookings Review*, 20 (2), 28–31.

——, 2004a, 'Putin, Yukos and Russia', *The Globalist*, 1 December 2004.

——, 2004b, 'Siberia: Russia's Economic Heartland and Daunting Dilemma', *Current History*, October.

——, 2004c, 'Eurasia on the Move', Presentation at the Kennan Institute, 27 September 2004.

——, 2004d, *Energy Empire: Oil, Gas and Russia's Revival*, The Foreign Policy Centre, September.

——, 2004e, 'Russia's Newly Found 'Soft Power'', *The Globalist*, 26 August 2004.

——, 2004f, 'China's Oil Strategy Not Conflicting with US Interest', Interview with *People's Daily Online*, 21 June 2004.

——, 2004g, 'Pipelines in the Caspian: Catalyst or Cure-all?' *Georgetown Journal of International Affairs*, Winter/Spring.

——, 2005, 'Governing Russia: Putin's Federal Dilemmas', *New Europe Review*, January.

Hill, F., & Fee, F., 2002, 'Fueling the Future: The Prospects for Russian Oil and Gas', *Demokratizatsiya*, 10 (4), 462–487, available at: http://www.demokratizatsiya.org

Hore-Lacey, I., 1999, 'Australia's faltering uranium revival', Paper presented at the 24th Annual Symposium of the Uranium Institute, London, September, available at: http://www.world-nuclear.org/sym/99idx.htm

Howell, D., 2005, 'Energy Myths and Illusions', *The Japan Times*, 15 August.

Huber, P. W., & Mills, M. P., 2005, *The Bottomless Well: The Twilight of Fuel: the Virtual of Waste, and Why We Will Never Run out of Energy*, Basic Books, New York.

Hydrocarbon Vision 2025, accessed on 8 April 2005, available at: http://www.petrodril.com/hydrocarbon.htm

IAGS (Institute of Analysis for Global Security), 2004, 'Chilly Response to US Plan To Deploy Forces in the Strait of Malacca', *Energy Security*, accessed on 24 May, available at: http://www.iags.org/n0524042.htm

IAGS (Institute of Analysis for Global Security), 2005, 'In Search of Crude China Goes to the Americas', accessed on 18 January, available at: http://www.iags.org/n0118041.htm

ICG, 2004, Repression and Regression in Turkmenistan: A New International Strategy, International Crisis Group, Asia Report, No. 85, accessed on 4 November, available at: http://www.crisisgroup.org/home/index.cfm?id=3091&CFID=7149152&CFTO KEN=78517829

IEA (International Energy Agency), 1999, *Coal in the Energy Supply of China*, OECD, Paris.

——, 2000, *China's Worldwide Quest for Energy Security*, OECD, Paris.

——, 2002, *Russia Energy Survey 2002*, OECD/IEA, Paris.

——, 2003a, *Energy Policies of IEA Countries: Japan 2003 Review*, OECD, Paris.

——, 2003b, *World Energy Investment Outlook 2003*, OECD, Paris.

——, 2003c, *Annual Statistical Supplement for 2002 (2003 Edition)*, OECD, Paris.

——, 2004a, *Analysis of the Impact of High Oil prices on the Global Economy*, OECD, Paris.

——, 2004b, *World Energy Outlook*, OECD, Paris.

——, 2004c, *Natural Gas Information*, 2004 Edition, OECD/IEA, Paris.

——, 2004d, *Key World Energy Statistics*, OECD/IEA, Paris.

——, 2005a, *Oil Market Report*, OECD/IEA, Paris, 11 April.

——, 2005b, *International Energy Outlook, 2005*, OECD/IEA, Paris.

——, 2005c, *Annual Energy Outlook, 2005*, OECD/IEA, Paris.

IET (Institute for the Economy in Transition), 2005a, Russian Economy in 2004: Trends and Perspectives, Issue 26, Moscow.

——, 2005b, Russian Economy in 2005: Trends and Perspectives, April, Moscow.

IISS (International Institute for Strategic Studies), 2004, 'Piracy and Maritime Terror in Southeast Asia', *Strategic Comments*, 10 (6), July, available at: http://www.iiss.org/stratcomfree.php?scID=386

Imber, M., 1989, *The USA, ILO, UNESCO and IAEA: Politicization and Withdrawal in the Specialised Agencies*, Macmillan, London.

IMF (International Monetary Fund), 2004, *World Economic Outlook 2004*, Washington DC.

——, 2005a, *Global Financial Market Development*, Washington DC.

——, 2005b, *World Economic Outlook 2005*, Washington DC.

——, 2005c, *Global Financial Stability Report*, Washington DC.

India–US Energy Dialogue Statement, 2005, Embassy of the United States of America, New Delhi, May 31.

Indyk, M., 1977, 'Australian Uranium and the Non-Proliferation Regime', *Australian Quarterly*, 49 (4).

Inglis, K. A. D., 1975, *Energy: From Surplus to Scarcity?*, Applied Science Publishers Ltd, Essex.

International Road Federation, *World Road Statistics*, 1998.

'Interview: Indian Prime Minister Singh', 2005, *Washington Post*, July 20.

Ivanov, V., & Hamada, M., 2002, 'Energy Security and Sustainable Development in Northeast Asia: prospects for US–Japan Coordination', Economic Research Institute for Northeast Asia and Institute of Economic Research, Russian Academy of Sciences, the Administration of Khabaroevskiy and the Korean Energy Economics Institute International Workshop, Khabarovsk, Russia, accessed on September 17–19, available at: http://www.erina.or.jp/En/Research/Energy/eEnergy.htm

Jaffe, A. M., 2001a, 'Asia's Thirst for Fuel: Cooperation and Energy Geopolitics', *Harvard International Review*, 23 (2), accessed 4 August 2005, available at: www.hir.harvard.edu/articles/906/1/

——, 2001b, 'World of Plenty: Energy as a Binding Factor,' *Erina Report*, 42, October 2001, Economic Research Institute for Northeast Asia, Japan, available at: http://www.erina.or.jp/En/Research/Energy/Jaffe42.pdf

Jaffe, A. M., & Lewis, S. W., 2002, 'Beijing's Oil Diplomacy', *Survival*, 44 (1), 115–134.

Jaffe, A. M., & Manning, R. A., 2001, 'Russia, Energy and the West', *Survival*, 43 (2).

Jaffe, A. M., & Soligo, R., 2004, 'The Future of Saudi Price Discrimination: The Effect of Russian Production', Rice University, The Energy Dimension in Russian Global Strategy: *The influence of Russian energy supply on pricing, security and oil geopolitics*, October.

Jervis, R., 2005, 'Why the Bush Doctrine Cannot Be Sustained', *Political Science Quarterly*, 120 (3), 351–377.

Jevons, W. S., 1906, in A. W. Flux (ed.) *The Coal Question: An Inquiry Concerning The Progress of the Nation And The Probable Exhaustion Of Our Coal Mines*, 3rd edn, rev., Macmillan, London.

Johnson, A. I., 2003, 'Is China a Status Quo Power?', *International Security*, 27 (4), 5–56.

Johnson's Russia List, 2005, Russia must break its dependence on energy exports, says Putin, from Interfax; JRL #9158, May 25, available at: http://www.cdi.org/russia/johnson/9152-6.cfm

Kahler, M., 1980, 'America's Foreign Economic Policy: Is the Old-Time Religion Good Enough?' *International Affairs*, 56 (3).

Kahn, J., 2005, 'Appetite for Energy Fuels China's Fear', *International Herald Tribune*, 28 June.

Kakuchi, 2005, 'Uncorking the Plutonium (energy) Genie', *Asia Times Online*, 10 May.

Kalam, A. P. J., 2005, *Energy Independence*, President of India, A. P. J Abdul Kalam's address to the nation on the eve of 59th Independence Day, accessed on 16 August 2005, available at: http://presidentofindia.nic.in/scripts/eventslatest1.jsp?=956

Karl, T. L., 1997, *The Paradox of Plenty: Oil Booms and Petro States*, University of California Press, Berkeley, CA.

Katsuhiko, H., 2000, 'China's Strategy for the Internationalisation of Energy Supplies and Asia's International Environment', in K. W. Radtke & R. Feddema (eds.) *Comprehensive Security in Asia: Views from Asia and the West on a Changing Security Environment*, Brill, Leiden.

Kelly, D., 2005, 'Rice, Oil and the Atom: A Study of the Role of Key Material Resources in the Security and Development of Japan', *Government and Opposition*, 40 (2), 278–327.

Kelly, J. A., 2004, 'Dealing with North Korea's Nuclear Program', Statement to the Senate Foreign Relations Committee, US Department of State, July 15, available at: http://www.state.gov/p/eap/rls/rm/2004/34395.htm

Keohane, R., 1982, 'International Agencies and the Art of the Possible: The Case of the IEA', *Journal of Policy Analysis and Management*, 1 (4), 471–474.

Kerin, J., 1988, *CPD* (House of Representatives), 20 April.

Kessler, C., 2004, *Nuclear Power's Growth in Asia Amid renewed Proliferation Concerns and Increasing Terrorism Threats*, prepared for a Conference on Asian Energy Security and Implications for the US, Seattle, September 28–29, available at: http://www.nbr.org/programs/energy/presentations/kessler.pdf

Khurana, G., 2005, *Securing maritime Lifelines: Is there a Sino–Indian 'Confluence'?*, Paper presented at an international seminar 'Sino–Indian Energy Cooperation' at The Center for South Asia–West China Cooperation & Development Studies (SAWCCAD), Sichuan University, Chengdu, China, 29 July.

Kim, M., 2005, 'Russian oil and gas: impacts on global supplies to 2020', *Australian Commodities*, 12 (2), 361–78, ABARE.

Kim, T., 2001, 'Security Reconsidered: An Eclectic Approach between Traditional and Human Security', *Flinders Journal of History and Politics*, 22, 2001/2002.

Klare, M. T., 2001a, 'The New Geography of Conflict', *Foreign Affairs*, June 2001.

——, 2001b, *Resource wars: the new landscape of global conflict*, Metropolitan Books, New York.

——, 2004a, *Blood and Oil: The Dangers and Consequences of America's Growing Oil Dependency*, Metropolitan Books, New York.

——, 2004b, 'Geopolitics Reborn: The Global Struggle over Oil and Gas Pipelines', *Current History*, December, pp. 428–433.

——, 2005a, 'Imperial Reach', *The Nation*, 25 April.

——, 2005b, 'The Intensifying Struggle for Global Energy', accessed on 18 August 2005, available at: http://www.tomdispatch.com/indexprint.mhtml?pid=2400

——, 2005c, CERA forecasts plentiful supply, *Petroleum Economist*, London, July 2005.

Kleveman, L. C., 2004a, 'Oil and the New Great Game', *The Nation magazine*, accessed on February 16, available at: http://www.newgreatgame.com/excerpts.htm#map

——, 2004b, *Oil and the New Great Game: Blood and Oil in Central Asia*, Atlantic Monthly Press, New York, available at: http://www.newgreatgame.com/

Kojima, A., 2005, 'East Asia's Thirst for Energy', *Japan Echo*, 32 (5), 32–35.

Koyama, K., 2001, 'Energy Security in Asia', Paper presented to the World Energy Council, 18th Congress, Buenos Aires, October, available at: http://www.worldenergy.org/wec-geis/publications/default/tech_papers/18th_Congress/ downloads/ ds/ds7/ds7_7.pdf

Krepinevich, A., 2002, *Defense Transformation*, in testimony before the US Armed Services Committee, April 9 2002, available at: http://www.csbaonline.org/cgi-local/pubfind.cgi?PubCategory=Transformation&no_top

Kupchan, C. A., (ed.), 2001, *Power in Transition: The Peaceful Change of International Order*, United Nations University Press, Tokyo.

Kuru, A., 2002, 'The Rentier State Model and Central Asian Studies: The Turkmen Case', *Alternatives*, 1 (1).

LaCasse, C., & Plourde, A., 1995, 'On the Renewal of Concern for the Security of Oil Supply', *Energy Journal*, 16 (2), 1–23.

Lahn, G., & Paik, K.-W., 2005, Russia's Oil and Gas Exports to North-East Asia, Sustainable Development Programme, Chatham House, April, available at: http://www.mideasti.org/pdfs/calabrese304.pdf, 10 St James's Square, London SW1Y 4LE, available at: www.chathamhouse.org.uk

Lajous, A., 2004, 'Production Management, Security of Demand and Market Stability', *Middle East Economic Survey*, 47 (39), accessed on September 27, available at: www.mees.com

Lambert, T., & Woollen, I., 2004, 'View of 12 million b/d Russian output by 2010 places focus on export limits', *Oil and Gas Journal Online*, PennWell, Tulsa, Oklahoma, 26 July.

Lan, X., 2003, 'Fruitful Investment Abroad', *Beijing Review*, 6 February.

——, 2004, 'A Nuclear Energetic Nation', *Beijing Review*, 10 June.

Larsson, R., 2006, *Russia's Energy Policy: Security Dimensions and Russia's Reliability as an Energy Supplier*, Division for Defence Analysis, Swedish Defence Research Agency (FOI), Stockholm.

Le Billon, P., 2004, 'The Geopolitical Economy of "Resource Wars"', *Geopolitics*, 9 (2004).

Lesbirel, S. H., 1998, *NIMBY Politics in Japan: Energy Siting and the Management of Environment Conflict*, Cornell University Press, Ithaca.

——, 2004, 'Diversification and Energy Security Risks: The Japanese Case', *Japanese Journal of Political Science*, 5 (1), 1–22.

Lewis, S. W., 2002, 'China's Oil Diplomacy and Relations with the Middle East', The James Baker III Institute for Public Policy, Rice University, September.

Li, D., 2003, 'China's Energy Challenged by the Pipeline Routes Dispute', *Economic Reference* (Jingji cankao bao), 12 August.

Li, Z., 2005, *Energy Security and Interaction: China vs. India,* Paper presented at an international seminar 'Sino–Indian Energy Cooperation' at The Center for South Asia–West China Cooperation & Development Studies (SAWCCAD), Sichuan University, Chengdu, China, 29 July.

Lieberthal, K., & Oksenberg, M., 1988, *Policy Making in China: Leaders, Structures, and Processes*, Princeton University Press, Princeton.

Lind, J. M., 2004, 'Pacificism or Passing the Buck? Testing Theories of Japanese Security Policy', *International Security*, 29 (1).

Logan, J., 2005, 'Energy Outlook for China: Focus on Oil and Gas', hearing on EIA Annual Energy Outlook for 2005, Committee on Energy and Natural Resources, US Senate, 3 February.

Lomborg, B., 2001, *The Sceptical Environmentalist: Measuring the Real State of the World*, Cambridge University Press, Cambridge and New York.

Luft, G., 2004, 'US, China Are on Collision Course Over Oil', *Los Angeles Times*, 2 February.

——, 2005, 'Reconstructing Iraq: Bringing Iraq's Economy Back Online', *Middle East Quarterly*, CNN.com, accessed on April 15 2004, available at: http://www.meforum.org/article/736

Luft, G., & Korin, A., 2004, 'The Sino–Saudi Connection', *Commentary Magazine*, (March).

Mabro, R., (ed.), 1980, *World Energy Issues and Policies*, Oxford University Press, New York.

——, 1998, 'OPEC Behaviour 1960–1998: A Review of the Literature', *Journal of Energy Literature*, IV (1).

——, 2005, 'The Reference Pricing System: Origins, Rationale, Assessment', Paper to EUROGULF, included in An EU–GCC Dialogue for Energy Stability and Sustainability: Final Report, April.

McIntosh–Baring, 1993, *Australian Coal Industry Study*, McIntosh & Co, Melbourne.

McKern B & Waltho, A. 1988, 'Australia', in Bruce McKern and Praipol Koomsup (eds), *Minerals Processing in the Industrialisation of ASEAN and Australia*, Allen and Unwin, Sydney.

Mackinder, H., 1950, 'The Geographical Pivot of History' in *Democratic Ideals and Reality*, Constable and Company, London.

McNeal, G. S., 2005, 'The Terrorist and the Grid', *The New York Times*, August 13.

Maddison, A., 1998, *China's Economic Performance in the Long Run*, OECD, Paris.

Mahan, A. T., 1890, *The Influence of Sea Power on History 1660–1783*, Sampson Low, Marston and Co. Ltd, London.

Malhotra, A. K., 2005, 'The Wolf at the Door', *Outlook*, 27 June.

Mandil, C., IAE Executive Director, 'Statement on High Oil Prices', IEA, Paris, July 2005.

Mangla, D., 2004, 'Transnational Piped Gas', *Power line*, February.

'Mani Shankar Aiyar inaugurates TERI Centre for Research on Energy Security' 2005, Press Release 31 May, accessed on 21 June 2005, available at: http://www.teriin.org/press_inside.php_id_10498

'Manmohan Cabinet Goes on Energy Overdrive Today', 2005, *Times of India*, August 11.

Manning, R. A., 2000a, *The Asian Energy Factor: Myths and Dilemmas of Energy, Security and the Pacific Future*, Palgrave, New York.

——, 2000b, 'The Asian Energy Predicament', *Survival*, 42 (3), 73–88.

Manning, R. A., & Davis, Z., 1997, 'PACATOM: Nuclear Cooperation in Asia', *The Washington Quarterly*, 20 (2).

Marcois, B. W., & Miller, L. R., 2005, 'China, US Interests Conflict', *Washington Times*, March 25.

Marks, R., 1986, 'Energy Issues and Policies in Australia', *Annual Review of Energy*, 11.

Marquardt, E., & Wolfe, A., 2005, Rice Attempts to Secure US Influence in Central Asia, PINR, accessed on 17 October, available at: http://www.pinr.com/report.php?ac=view_report&report_id=382&language_id=1

May, M., 1998, 'Energy and Security in East Asia', Asia/Pacific Research Centre.

Mead, W. R., 2004, *Power, Terror, Peace and War: America's Grand Strategy in a World at Risk*, Alfred A Knopf, New York.

Ministry of External Affairs, 2002, 'Southeast Asia and the Pacific', *Annual Report 2001–2002*, Government of India.

——, 2003a, *Joint Declaration on the Framework of Comprehensive Cooperation between the Republic of India and the Socialist Republic of Vietnam as they Enter the 21st Century*, Government of India, accessed on August 3, 2005, available at: http://meaindia.nic.in/

——, 2003b, *Memorandum of Understanding for Enhanced Cooperation in the field of Renewable Energy between the Ministry of Non-Conventional Energy Sources, Government of the Republic of India and the Ministry of Water Resources, Government of People's Republic of China*, Government of India, 23 June, accessed on August 5 2005, available at: http://meaindia.nic.in/

——, 2005a, *India–Vietnam Relations*, Government of India, accessed on 3 August 2005, available at: http://meaindia.nic.in/

——, 2005b, *Joint Press Statement* (Trilateral-between Minister for Energy of Myanmar, Minister of Energy and Mineral Resources of Bangladesh and Minster of Petroleum and Natural Gas of India), Government of India, accessed on 17 August 2005, available at: http://meaindia.nic.in/pressrelease/2005/01/13js02.htm

——, 2005c, *Joint Statement, India–Japan Partnership in a New Asian Era: Strategic Orientation of India–Japan Global Partnership*, Government of India, 29 April, accessed on 5 August 2005, available at: http://meaindia.nic.in/sshome.htm

Ministry of Finance, 2000, *Economic Survey 2000–01*, Government of India, accessed on 4 July 2005, available at: http://indiabudget.nic.in/es2000-01/chap77.pdf

Ministry of Petroleum and Natural Gas, 2004, *Annual Report 2003–04*, Government of India.

Misra, A., 2005, 'Contours of India's Energy Security: Harmonising Domestic and External Options', Draft paper presented to Workshop, *Energy Security in the Asia–Pacific Region* hosted by the Griffith Asia Institute and the Asia–Pacific Futures Network; Stamford Plaza Hotel, Brisbane, 31 August and 2 September.

Mitchell, J. V., 1996, *The New Geopolitics of Energy*, The Royal Institute of International Affairs, London.

MNES (Ministry of Non-Conventional Energy Sources), 2005, 'Overview', *Annual Report 2004–05*.

Morse, E. L., 1983, 'The International Petroleum Economy: A Rebirth of Liberalism?', *SAIS Review*, 3 (2).

——, 2004, 'The US and the Changing Geopolitics of Asian Energy: Security and Foreign Policy Interests', NBR and Asian Security Conference, Seattle, 29 September, available at: (http://www.nbr.org/energy/aesconference.html)

Morse, E. L., & Richard, J., 2002, 'The Battle for Energy Dominance', *Foreign Affairs*, March/April, 16–31.

Muni, S. D., & Pant, G., 2005, *India's Search for Energy Security*, Rupa &Co, New Delhi.

Murray, R., 1972, *Fuels Rush In: Oil and Gas in Australia*, Macmillan, Melbourne.

'National Energy Leading Group Set Up', 2005, 7 June, available at: http://www1.cei.gov.cn/ce/doc/cep5/200506071325.htm

Naughton, B., 1995, *Growing Out of the Plan: Chinese Economic Reform, 1978–1993*, Cambridge University Press, New York.

NBR (National Bureau of Asian Research), 2004, 'Conference Executive Summary' Asia Energy Security and Implications for the US', accessed on 28–29 September, available at: http://www.nbr.org/programs/energy/aescexec.pdf

NBS (National Bureau of Statistics of China), 2004, *China Statistical Yearbook, 2004*, China Statistics Press, Beijing.

Neff, A., 2005a, 'Caspian nations pursuing oil exports at greatly varying paces', *Oil & Gas Journal Tulsa*, 103 (22), 34–39.

——, 2005b, 'Caspian oil not likely to fill market void or depress prices', *Oil & Gas Journal*, 103 (21), 39–43.

Neff, T. L., 1984, *The International Uranium Market*, Ballinger, Cambridge MA.

——, 2005, 'Legacies form the future: the history of uranium', *Nuclear Engineering International*, 50 (606).

Negishi, M., 2005, 'Teikoku Oil gets drilling rights in East China Sea', *Japan Times online*, accessed on 15 July 2005.

Newbery, D. M., 1999, *Privatization, Restructuring, and Regulation of Network Utilities*, The MIT Press, Cambridge, Mass.

NIC (National Intelligence Council), 2004, *Mapping the Global Future*, United States Government Printing Office, Washington DC.

NIDS (National Institute for Defense Studies), 2004, *East Asian Strategic Review*, Japan Times, Tokyo.

Nishimura, H., 2005, 'Yushutsukoku ni ashimoto o mirareru Nihon no LNG ch?tatsu (Supply of LNG in Japan)', *Foresight* (Japan), March 2005, 72–73.

Nolt, J. H., 2002, 'China's Declining Military Power', *The Brown Journal of World Affairs*, 9 (1), available at: http://www.watsoninstitute.org/bjwa/archive/9.1/Essays/Nolt.pdf

Odell, P. R., 1997, 'The Global Oil Industry: The Location of Production – Middle East Domination or Regionalisation?' *Regional Studies*, 31 (3), 311–322.

OECD (Organisation for Economic Cooperation and Development), 2004, *Economic Survey of the Russian Federation 2004*, OECD, Paris, available at: http://www.oecd.org/dataoecd/56/0/32389025.pdf

Office of the Secretary of Defense, 2005, *The Military Power of the People's Republic of China 2005*, USGPO, Washington DC.

Ogawa, Y., 2004, 'The Asian Premium and Oil and Gas Supply from Russia', Rice University, The Energy Dimension in Russian Global Strategy: *The influence of Russian energy supply on pricing, security and oil geopolitics*, October.

Ogilvie-White, T., 2005, 'Preventing Nuclear and Radiological Terrorism: Nuclear Security in Southeast Asia', The Australian Centre for Peace and Conflict Studies Occasional Paper Series, available at: http://www.uq.edu.au/acpacs/docs/papers/NuclearSecurityOgilvie-White.pdf

Ogutcu, M., 2003, 'China's Energy Security: Geopolitical Implications for Asia and Beyond', *Oil Gas and Energy Law Intelligence*, 1 (2).

'Oil Money Divides Nigeria', 2005, National Public Radio, accessed on August 23 2005, available at: http://www.npr.org/templates/story/story.php?storyId=4803867

'ONGC Eyes Alternate Energy Sources', 2005, *Times of India*, 16 August.

OPEC (Organization of the Petroleum Exporting Countries), *Oil Outlook to 2025*, Vienna, 2004.

Osmundsen, P., Mohn, K., Misund, B., & Asche, F., 2006, 'Is oil supply choked by financial market pressures?', *Energy Policy*, accessed on 24 January 2006, available online.

Oxford Analytica, 2004, Russia: Rivals may divert Siberian energy exports, *Oxford Analytica 1997–2005*, January 15, available at: http://www.oxanstore.com/displayfree.php?NewsItemID=97161

Pachauri, R. K., 'Need for Integrated Policy', accessed on 21 June 2005, available at: http://www.teriin.org/features/art28.htm

'Pakistan, India play pipeline tune', 2005, *Indian Express*, 14 July, accessed on 3 August 2005, available at: http://www.indianexpress.com/res/web/pIe/full_story.php?content_id=74373

Palast, G., 2005, 'OPEC on the march', *Harper's Magazine*, New York, April, 310 (1859).

Paris, R., 2001, 'Human Security, Paradigm Shift or Hot Air?', *International Security*, 26 (2).

Parthasarathy, G., & Kurian, N., 2002, 'Enhancing India's Energy Security Options', *Pacific and Asian Journal of Energy*, 12 (1).

Pdgaard, P., 2002, *Maritime Security between China and Southeast Asia: Conflict and Cooperation in the Making of Regional Order*, Ashgate, Aldershot.

Pei, M., 2005, at a roundtable discussion on 'China and Geopolitics of Energy: UNOCAL and Beyond', Centre for American Progress, July 20, available at: http://www.american progress.org

Peng, L., 1997, 'Li Peng's China Energy Speech', available at: http://www.pnl.gov/china/lipend.htm

People's Daily, 2005, China starts work on Kazakhstan–China oil pipeline, People's Daily Online, March 25, available at: http://english.people.com.cn/200503/25/eng20050325 _178162.html

Percival, B., 2005, 'Indonesia and the United States: Shared Interests in Maritime Security', published by the United States–Indonesia Society, June, available at: http://www.usindo.org/pdf/Maritime%20Security.pdf

Peters, D., 2005, 'Carbon trading a long way off for Aust – Campbell', *Australian Associated Press General News*, 10 August.

Peters, S., 2004, 'Coercive Western Energy Security Strategies: "Resource Wars" as a New threat to Global Security', *Geopolitics*, 9 (1).

Petroleum Economist, 2005, China: Fuel shortages prompt re-think on oil pricing, *Petroleum Economist*, October, London.

Pindyck, R. S., 1991, 'Irreversibility, Uncertainty and Investment', *Journal of Economic Literature*, XXIX, 1110–1148.

Pinkus, B., 2002, 'Atomic Power to Israel's rescue: French–Israeli nuclear cooperation, 1945–1957', *Israel Studies*, 7 (1).

'PM Draws Up Energy Blueprint', 2005, *Times of India*, 7 August 2005.

Porter, E. D., 2001, 'US Energy Policy, Economic Sanctions and World Oil Supply', American Petroleum Institute, Washington DC.

Pravda, R. U., 2005, 'Russia to become strategic economic partner of the European Union', *Pravda RU*, 4 October, available at: http://www.worldpress.org/link.cfm? http://english.pravda.ru/main/18/88/354/16245_EU.html

Przeworski, A., & Limongi, F., 1997, 'Modernization: Theories and Facts', *World Politics*, 49 (January).

Pultz, C. B., 2003, 'The PLA and China's Changing Security Environment', *Security Insight*, II:1, January, available at: http://www.ccc.nps.navy.mil/si/jan03/eastAsia.asp

Pustilnik, M., 2005, Winners and Losers of Gazprom–Rosneft Merger, *MosNews*, 3 March.

Raman, B., 2005, 'The Hype Behind India's Japan Ties', *Asia Times,* accessed on 12 August 2005, available at: http:atimes.com/atimes/South_Asia/GF16Df02.html

Rashid, A., 2001, *Taliban: The Story of the Afghan Wars*, Part 3, Pan Books, London.

Renner, M., 2003, 'Post-Saddam Iraq: Linchpin of a New Oil Order', *FPIF Policy Report,* (citing Energy Intelligence Group), available at: http://www.fpif.org/papers/oil_body.html)

Research and Information System for Developing Countries, 2004, 'Towards an Asian Energy Community: An Exploration', High-Level Conference on Asian Economic Integration: Vision of a New Asia conference, Tokyo, 18–19 November.

Reuters, 2005a, 'Goldman Sachs: Oil Could Spike to $105', *Energy Bulletin*, 31 March.

Reuters, 2005b, 'Russia looks for way to become an oil superpower', October 2005, available at: http://www.financialexpress.com/latest_full_story.php?content_id=100547

Roberts, P., 2004, *The End of Oil: On the Edge of a Perilous New World*, Houghton Mifflin Company, Boston.

Rodrigue, J. P., 2004, 'Straits, Passages and Chokepoints: A Maritime Geostrategy of Petroleum Distribution', *Les Cahiers de Geographie du Quebec*, special issue on strategic straits, 48 (135).

Romeo, S., 2004, 'China is Emerging as a Rival to US for Oil in Canada', *The New York Times*, 23 December.

Ross, M. L., 2001, 'Does oil hinder democracy?', *World Politics*, 53, 325–361.

——, 2004a, 'How does natural resource wealth influence civil wars? Evidence from thirteen cases', *International Organization*, 58 (1), 35–67.

——, 2004b, 'What do we know about natural resources and civil war?', *Journal of Peace Research*, 41 (3), 337–356.

'Russia Snubs China over Pipeline Route', 2004, *Times on Line*, accessed on September 24, available at: http://www.timesonline.co.uk/article/0,,5-1278745,00.html

Sachs, J. D., & Warner, A. M., 1999, 'The big push, natural resource booms and growth', *Journal of Development Economics*, 59, 43–76.

Saddler, H., & Ulph, A., 1980, 'The Australian Liquid Fuel Situation', in T. Van Dugteren (ed.) *Oil and Australia's Future*, Proceedings of the 45th Summer School of the Australian Institute of Political Science, Hodder and Stoughton, Sydney.

Sakamoto, Y., 2005, 'The balance of power', interview, *The Japan Journal*, January, 24–25.

Salameh, M. G., 2001, 'A Third Oil Crisis?', *Survival*, 43 (3).

——, 2003, 'Quest for Middle East Oil: The US Versus the Asia–Pacific Region', *Energy Policy*, 31.

Scarrow, H. A., 1972, 'The Impact of British Domestic Air Pollution Legislation', *British Journal of Political Science*, 2, 261–282.

Scherer, F. M., 1980, *Industrial Market Structure and Economic Performance*, 2nd edn, Houghton Mifflin.

Schlesinger, J., 2005, 'The Theology of Global Warming', *The Asian Wall Street Journal*, 10 August.

Schneider, M., & Froggatt, A., 2004, *The World Nuclear Industry Status Report 2004*, commissioned for the Greens–EFA Group in the European Parliament, Brussels, December.

Schumpeter, J. A., 1943–1966, *Capitalism, Socialism and Democracy*, Unwin University Books.

Schwarz, B., & Layne, C., 2002, 'A new grand strategy', *The Atlantic Monthly* Boston, 289 (1), 36 (citing W. R. Mead in an interview on National Public Radio early in October 2001).

SCO (Shanghai Cooperation Organisation), 2006, *Chronology of Main events within the framework of "Shanghai five" and Shanghai Cooperation organization (SCO)*, accessed on 30 March 2006, available at: http://www.sectsco.org/html/00030.html

Scott, R., 1994, *The History of the IEA: Origins and Structure*, International Energy Agency, Paris.

SEEN (Sustainable Energy and Economy Network), 2004, *Project Profile: Baku–Tblisi–Ceyhan Oil Pipeline*, Institute for Policy Studies, available at: http://www.seen.org/db/Dispatch?action-ProjectWidget:59-detail=1

Shambaugh, D., 2005, 'The New Strategic Triangle: US and European Reactions to China's Rise', *The Washington Quarterly*, 28 (3), 7–25.

Shin, E. S., 2005, *Joint Stockpiling and Emergency Sharing of Oil: Updates on the Situations in the ROK and Arrangements for Regional Cooperation in Northeast Asia*, Asian Energy Security Workshop, May 13–16, available at: http://66.102.7.104/

search?q=cache:xUthkRX5pNsJ:www.nautilus.org/aesnet/2005/JUN2205/Shin_Stock
pile.ppt+Japan+and+Korea+strategic+petroleum+reserve&hl=en

Shlaim, A., 1997, 'The protocol of Sèvres, 1956: anatomy of a war plot', *International Affairs*, 73 (3).

Shleifer, A., & Treisman, D., 2004, 'A Normal Country', *Foreign* Affairs, New York, March/April, 83 (2), 20.

Simmons, M., 2005, *Twilight in the Desert: The Coming Saudi Oil Shock and the World Economy*, John Wiley & Sons, New Jersey.

Singh, M., 2004, 'Economic Utilization of Energy', Prime Minister's address on National Energy Conservation Day, *Power Line*, December.

Singh, S., 2005, *Address by External Affairs Minister Shri K. Natwar Singh, at the Inauguration of the India–Japan Parliamentary Forum, Federation House, New Delhi*, Ministry of External Affairs, Government of India, 29 April, accessed on 7 August 2005, available at: http://meaindia.nic.in/sshome.htm

Sinton, J., & Fridley, D. G., 2000, 'What Goes up: Recent Trends in China's Energy Consumption', *Energy Policy*, 28, 671–687.

Skinner, R., 2005, 'Crude Oil: Scenarios and Perspectives of the Market', Presentation, August, Comisión de Investigación de los Precios del Petróleo, Querétaro, Mexico, available at: http://www.oxfordenergy.org/presentations/CrudeOilScenarios.pdf

Skinner, R., & Arnott, R., 2005a, *EUROGULF: An EU-GCC Dialogue for Energy Stability and Sustainability*, accessed on June 4 2005, available at: http://Europa.eu.int/comm/energy/index_en.html

——, 2005b, *The Oil Supply and Demand Context for the Security of oil Supply to the EU from the GCC Countries*, Oxford Institute for Energy Studies, April.

Smil, V., & Knowland, W. E., 1980, *Energy in the Developing World*, Oxford University Press, New York.

Smith, T., 1979, 'Forming a uranium policy: why the controversy?', *Australian Quarterly*, 51 (4).

Snyder, J. C., 1985, 'Iraq', in J. C. Snyder & S. F. Wells, Jr. (eds.) *Limiting Nuclear Proliferation*, Ballinger, Cambridge.

Sproull, R. 1997, 'WMC plans $900m mine upgrade', *The Australian*, 25th April.

Starick, P., 2004, 'Alexander the survivor', *The Advertiser*, 20 December.

Stobaugh, R., & Yergin, D., 1979, *Energy Future*, Random House, New York.

Story, J., 2004, 'The Global Implications of China's Thirst for Energy', *Middle East Economic Survey (MEES)*, 16 February, XLVII (7).

Strausz-Hupe, R., 1942, *Geopolitics: The Struggle for Space and Power*, G. P. Putnam's Sons, New York.

Sykes, T., 1978, *The Money Miners: The Great Australian Mining Boom*, Fontana/Collins, Melbourne.

Szyliowicz, J. S. & O'Neill, B. E., 1975, *The Energy Crisis and US Foreign Policy*, Praeger Publishers, New York.

Tang, Y., 2003, 'China's Oil Industry: Coping with Risks', *Beijing Review*, 10 April.

Telhami, S., & Hill, F., 2002, 'Does Saudi Arabia Still Matter? Differing Perspectives on the Kingdom and Its Oil America's Vital Stakes in Saudi Arabia', *Foreign Affairs*, Nov/Dec, 81 (6).

'The Afghan Corridor: Prospects for Pakistan–Central Asia Relations in Post-Taliban Afghanistan?', 2002, *Spotlight on Regional Affairs*, September, accessed on 2 August 2005, available at: http://irs.org.pk/spotlightEditions/sept2002.pdf

Till, G., 2001, 'A Changing Focus for the Protection of Shipping', *Journal of Indian Ocean Studies*, April 1, 9 (1).

Toichi, T., 2003, 'Energy Security in Asia and Japanese Policy', accessed on 18 August 2005, available at: www.eneken.ieej.or.jp/en/data.pdf.200/pdf

——, 2005, 'Enerugi gaik? de sonzaikan o masu Indo' (India raises its profile via energy diplomacy), *Foresight* (Japan), March, 74–75.

Tonnesson, S., 2000, 'China and the South China Sea: A Peaceful Proposal', *Security Dialogue*, 31 (3), 307–326.

Troush, S., 1999, 'China's Changing Oil Strategy and its Foreign Policy Implications', *CNPS Working Paper*, The Brookings Institutions.

ul Haque, I., 2005, 'Tri-nation Pipeline Makes Headway: Foreign Financing Likely', *Dawn Internet Edition,* accessed on 29 July 2005, available at: http://www.dawn.com/2005/07/29/top1.htm

Umbach, F., 2004, 'Global Energy Supply and Geopolitical Challenges' in F. Godement, F. Nicolas, & T. Yakushiji (eds.) *Asia and Europe. Cooperating for Energy Security*, A Council for Asia–Europe Cooperation (CAEC) – Task Force Report, Paris.

UNCTAD, (United Nations Conference on Trade and Development), 2004, *World Investment Report, 2004*, UNCTAD, Geneva.

UNDP (United Nations Development Programme), 1994, *Human Development Report 1994*, Oxford University Press, New York.

——, 2000, *UNDP Millennium Development Goals*, available at: http://www.un.org/millenniumgoals

United States Department of Energy, 2006, *National Security Review of International Energy Requirements*, Report to Congress under Section 1837 of the Energy Policy Act 2005, February.

United States House of Representatives, 1976, Committee on Interstate and Foreign Commerce, Sub-Committee on Oversight and Investigations, *International Uranium Supply and Demand*, 94th Congress, 4 November.

'US warns India over Iran stance', 2006, *BBC,* 25 January, available at: http://news.bbc.co.uk/2/hi/south_asia/4647956.stm

USA Today, 2004, 'China enacts first fuel-efficiency standards', 8 October, available at: http://www.usatoday.com/money/world/2004-10-08-china-fuel-efficiency_x.htm

Uyama, T., 2003, Japanese Policies in Relation to Kazakhstan: Is There a "Strategy"? in R. Legvold (ed.) *Thinking Strategically: The Major Powers, Kazakhstan, and the Central Asian Nexus*, MIT Press.

Vatansever, A., 2005, *Russia's Growing Dependence on Oil and Its Venture into a Stabilization Fund*, The Institute for the Analysis of Global Security (IAGS), March 28, available at: http://www.iags.org/n0328052.htm

Viswanathan, R., 2005, 'India's Energy quest in Latin America', *The Hindu*, 31 March 2005.

Wang, J., 2004a, 'China's Changing Role in Asia', Institute of America Studies, Chinese Academy of Social Sciences, January.

Wang, Y., 2004b, 'On the Rise', *Beijing Review*, June 17 2004.

Wang, Y., Gu, A., & Zhang, A., 2005, 'Updates on the Chinese Energy Sector and the China LEAP Model', 3rd Asia Energy Security Workshop, 13–16 May, Beijing, China, available at: http://www.nautilus.org/energy/2005/beijingworkshop/index.html

Washingtonpost.com, 2005, 'Indonesia May Cut Fuel Subsidies', The Associated Press, September 8, available at: http://www.washingtonpost.com/wp-dyn/content/article/2005/09/08/AR2005090800113_pf.html

Watkins, E., 2005a, 'Progress sought on Siberia–Pacific oil line', *Oil & Gas Journal*, October 24.

——, 2005b, 'Russian oil exports by rail to China rising', *Oil & Gas Journal*, October 18, available at: http://ogj.pennnet.com/articles/article_display.cfm?article_id=239336

Wayne, E. A., 2005, Assistant Secretary of State for Economic and Business Affairs, Testimony to the US Senate Committee on Foreign Relations, 26 July.

Weinstein, M., 2005a, Intelligence Brief: Uzbekistan-C.I.S. PINR, 1 September, available at: http://www.pinr.com/report.php?ac=view_report&report_id=358

——, 2005b, 'Intelligence Brief: Shanghai Cooperation Organization', *The Power and Interest News Report* (PINR), 12 July, available at: http://www.pinr.com/report.php?ac=view_report&report_id=325&language_id=1

Wen, J., 2005, Chinese Premier, at the first meeting of the Leading Energy Group, "Energy Leading Group Set Up," *China Daily*, 4 June.

Wolfe, A., 2005, 'The "Great Game" Heats Up in Central Asia', *The Power and Interest News Report* (PINR), 3 August, available at: http://www.pinr.com/report.php?ac=view_report&report_id=339&language_id=1

Woodrow, T., 2002, 'The Sino-Saudi Connection', *China Brief*, 2 (21), October 24.

Woodward, K., 1980, *The International Energy Relations of China*, Stanford University Press, Stanford, CA.

World Bank, 2005, *Global Economic Prospects 2005,* Washington DC, Appendix: Regional Economic Prospects, Table A1.

Wu, K., & Han, S. L., 2005, 'Chinese Companies Pursue Overseas Oil and Gas Assets', *Oil and Gas Journal*, 103 (15), 18–25.

Xu, X., 1999, 'The Oil and Gas Links between Central Asia and China: A Geopolitical Perspective', *OPEC Review*, March, 33–54.

Xu, Y., 2002, *Powering China*, Ashgate, Dartmouth.

——, 2005, China's Energy Security, Draft paper presented to Workshop, *Energy Security in the Asia–Pacific Region* hosted by the Griffith Asia Institute and the Asia–Pacific Futures Network; Stamford Plaza Hotel, Brisbane, 31 August and 2 September.

Yenukov, M., 2005, 'Russian Pacific Pipeline Hits Environment Trouble', *Planet Ark*, 16 June, available at: http://www.planetark.com/dailynewsstory.cfm/newsid/31260/newsDate/16-Jun-2005/story.htm

Yergin, D., & Stanislaw, J., 1998, *The Commanding Heights: The Battle Between Government and the Marketplace That is Remaking the Modern World*, Simon & Schuster, New York.

The Yomiuri Shinbun, 2005a, 'Planning National Strategies: Resources and Energy/China Fomenting Future Future Crisis', *The Daily Yomiuri Online*, accessed on 14 April 2005.

——, 2005b, 'Japan to Get Oil via Siberia Pipeline in "12"', *The Daily Yomiuri Online*, accessed on 2 May 2005.

——, 2005c, 'Planning National Strategies: Paper Fuels N-energy Debate', *The Daily Yomiuri Online*, 5 May, accessed on 10 May 2005.

Zanoyan, V., 2004, 'The Oil Investment Climate', *Middle East Economic Survey*, xlviii (26), 28 June.

Zhai, B., 2003, 'Stepping Up Oil Reserves', *Beijing Review*, 20 February.

Zheng, Y., 2004, 'Being Diplomatic', *Beijing Review*, July 22.

Zweig, D., & Bi, J., 2005, 'China's Global Hunt for Energy', *Foreign Affairs*, September/October.

Index

For Product Safety Concerns and Information please contact our EU
representative GPSR@taylorandfrancis.com
Taylor & Francis Verlag GmbH, Kaufingerstraße 24, 80331 München, Germany

www.ingramcontent.com/pod-product-compliance
Lightning Source LLC
Chambersburg PA
CBHW071848270326
41929CB00013B/2149

* 9 7 8 0 4 1 5 6 4 7 4 8 9 *